SMALL GAME HUNTER

PETER SMITHERS

BRAMBLEBY BOOKS

Small Game Hunter
Text and Photos © Peter Smithers, 2024

All Rights Reserved. No part of this book may be reproduced in any form by photocopying or by any electronic or mechanical means, including information, storage or retrieval systems, without permission in writing from both the copyright owner and the publisher of this book.

Peter Smithers has asserted his right under the Copyright Design And Patent Act, 1988, to be identified as author of this work.

A CIP catalogue record for this book is available from the British Library

ISBN 9781908241702
eISBN 9781908241757

Cover design by Pascale Dilger
Book layout by Lapiz Digital Services, India

Published 2024 by Brambleby Books Ltd.
www.bramblebybooks.co.uk

Printed by Halstan Printing Group, Amersham, England

Contents

A Song for Now — iv
Preface — vii
Acknowledgements — ix

Chapter 1: Of light and scale – Lepidoptera (the moths and butterflies) — 1
Chapter 2: Aphrodite Rising – dragons, damsels and river flies — 13
Chapter 3: Spiders — 23
Chapter 4: Diptera — 59
Chapter 5: Beetles, grasshoppers, crickets and other things — 73
Chapter 6: Insects, a new hope — 87
Chapter 7: How one curious thing leads to another even more curious thing: bug hunts and exhibitions — 109
Chapter 8: Art and entomology: two cultures, both alike in dignity, in fair academia where we lay our scene — 145
Chapter 9: Telson (the last segment of an invertebrate's body) — 165

About the Author — 171

A Song for Now

I heard it on the radio
they said that time is running out
the silent spring they warned us of
is almost here, it's not in doubt

I saw it when I walked the dog
the butterflies and birds were few
the grasshoppers, they failed to sing
the crickets they had vanished too

I read it in a magazine
the next extinction has begun
we're losing species every day
they're going nowhere one by one

I saw it on the TV news
that droughts are long and frequent now
that floods are spreading far and wide
our forest burn and take the trees
the trees we need to save, but how

I saw it in a documentary
we need to start a green new deal
to change the way we run the world
to work with nature, not to steal

I heard it often from MPs
We've ran down oil and planted trees
electric cars will give clean air
they're working hard its coming soon
but they do have voters to appease

I hear it from my grandson now
are the swallows coming back
why are butterflies so rare
and were the hedgehogs always flat

I hear schoolchildren chant and shout
their angry march, "the worlds on fire"
change must come now, there is no time
for compromise and blah blah blah

it seems we're standing on the track
the train is coming down the line
yet we discuss if trains are real
if moving off is really wise
or if we stay we could be fine

We need to act, it's down to us
it's time to change the way we live
to live with nature once again
to nurture and not just consume
the future watchword is SUSTAIN

Peter Smithers

Preface

Metamorphosis

When I was at school I was fascinated by words. My English teacher, the formidable Mrs. Gulliver, always ended her English lessons with a session on the study of words, and I would eagerly await these brief sessions of etymology. I loved the origins and histories of words, their subtleties of meaning, the way they sound. I was enthralled by their complexity and their power to paint images. Words were magical in every respect, except how they were spelt. This particular aspect would often elude me. One day, in a moment of exasperation, Mrs. Gulliver turned to me and said "Smithers, if you don't improve your spelling, you will end up studying insects rather than words." Prophetic or what? I wish I could have told her that she was right and that it had been a good choice, one that shaped my life.

The invertebrate world is a truly amazing realm that teams with life. It is bizarre beyond credibility but is also stunningly beautiful, and this strange beauty was the magnet that drew me in. It's a world the scale of which lies just below the level of our human perception, so it passes us by, unnoticed, unappreciated and uncared for. But if you stop and look it can be a revelation.

In the early 90s the University of Plymouth (formerly Plymouth Polytechnic until 1992, when it gained university status) would take first-year students to Swanage, Dorset, for a week of ecology field work. One of the studies was to examine chalk downland vegetation, and the first thing we would ask them to do was to lay on the ground with their noses just above the turf. They would push the plants aside to make a small open space and spend the next ten minutes watching what walked across it. They were always amazed at the procession of invertebrates that ambled across. Looking with the naked eye can be exciting, but to get a closer look at the invertebrate world, a hand lens or microscope is required. These will then take you down to the scale at which they live.

Robert Hooke's book *Micrographia*, published in 1665, revealed this microscopic world for the first time to a stunned 17th-century London. It contained detailed drawings of flies, fleas, head lice, as well as pins and strands of cloth that he had made using a self-built microscope, the first one in Britain. It generated a wave of horror and revulsion across London society but also ignited a public curiosity that is still with us today. The hugely popular photographic exhibition *Microsculpture* by Levin Biss (2017), which displays large photographs of very small insects, is testament to this. The entomologist and conservationist Miriam Rothschild (1908-2005) declared "Once you have a microscope, life is never long enough." I can personally acknowledge that this is, oh, so true. A fascination with the microcosm is addictive, and I was hooked as soon as I arrived at Plymouth Polytechnic back in 1974. They had just received a batch of new stereo zoom binocular microscopes, the first such microscope that I had ever seen. Suddenly the microscopic world became crystal clear, no longer fuzzy images squinted at with one eye that I was used to with the old monocular microscopes that I had access to at my previous job. Here was a bright, three-dimensional image that brought this micro-world into vivid focus. This revelation activated a curiosity that my father had planted in my head many years before. After Sunday afternoon walks with the family and friends, we would retire to the sitting room with his clutch of *Observers Books* and work out which plants and butterflies we had seen that day. These were early moments of wonder as the books introduced me to the names and lives of the curious insects that glided past us in those now vanished Surrey woodlands. Checking out the names of the butterflies and hedgerow plants was as far as we could go, yet it planted that seed of curiosity in my head, a curiosity about the natural world that sprang back to life when I realised I had access to the heart of the microcosm.

Acknowledgements

Small Game Hunter began as a set of notes for my children, but it has grown into a memoir of my encounters with invertebrates and an appreciation of the amazing world that we share with so many other living organisms. I hope it will inspire readers to stop and look, pause and listen, linger and appreciate the vast array of life that surrounds us.

It has had a genesis of around twelve years, over which time many people have lent their support and inspiration. I would like to thank my wife Virginia who had to endure reading all the very rough first drafts and, despite this, encouraged me to continue! Also, Jane Lamerton and Margaret Matthews who read the second drafts, made suggestions and demanded that I keep writing. Without the support of these three ladies those first drafts would have remained buried at the back of my computer hard drive.

I also owe much to my colleagues at the University of Plymouth who provided the many opportunities for me to meet and study the huge diversity of insect life that surrounds us. In particular Mick Uttley and Ken Thompson who dragged me from the comfort of Devon to the wild lands of northern Britain and who also introduced me to the beauty and complexity of our flora. And to Steve Burchett, John Bull and Steve Hill who invited me to South-East Asia where I became intoxicated by the biodiversity of the tropics. Thank you to all of you for your support and encouragement.

Lastly, my sincere thanks to Nicola and Hugh Loxdale at Brambleby Books Ltd., who continually encouraged me to finish the task I had begun so long ago and have made *Small Game Hunter* a reality.

Chapter 1

Of light and scale – Lepidoptera (the moths and butterflies)

The phrase 'like moths to a flame' is one we all understand. Most of us have experienced insects buzzing around lights in the garden or can remember those summer evenings when moths and other insects bounce up against our lighted windows, colliding against the glass in a desperate dance to get inside, a dance that can either delight or dismay those indoors. Streetlamps will also attract flying insects on warm spring and summer evenings, a cloud of tiny specks that whirl in and out of the pools of light that hang high above our streets, a flight of tiny wings lost in the glare of our modern technology. But why does this happen, and what draws all those insects to the bright lights of human habitation? The answer is that we are not really sure.

There have been several theories, but none of them have been proven. One was that moths navigate by flying at a constant angle to the moon. As the moon is a long way off when moths do this, the moth would fly in a straight line. The advent of campfires, candles, then gas lighting and finally electric light offered many choices of bright lights to lock onto and the moths were confused. Flying at a constant angle to a nearby light means the moth would fly in an ever-decreasing circle. However, no one has ever clearly demonstrated that this is the case.

Another idea was that the chemicals that female moths emit to attract males (pheromones) reflect infrared light, so maybe artificial lights were generating just this frequency and fooling male moths into chasing after phantom females. But male and female moths are equally attracted to artificial light, so that idea was thrown out. What we do know is that the light sources that emit light in the ultraviolet (UV) range of the spectra attract more moths, and the smaller and brighter the light source the greater the attraction. This is of course modified by several factors, such as temperature (most insects need warm nights to be able to fly), time of year (if a moth is in its larval stage it can't fly at all) and the phase of the moon (if the moon is bright, attraction to artificial lights is reduced). The reasons for these behaviours remain a mystery.

Relying on lighted windows and streetlights is one way to hunt for and collect moths, but entomologists who study them nowadays tend to use a device known as a light trap. These utilise the fact that very small bright lights that emit UV light attract large numbers of moths. For this reason, small but bright lights are mounted over a large collecting container that allows the moths to fall in and be retained. These traps are amazing; place one anywhere in the English countryside and a vast array of flying insects will loom out of the surrounding darkness, queuing up to get to the light. I have spent many nights sitting on the ground next to such a light trap watching as moth after moth spiralled out of the darkness into the intense circle of light that surrounds the trap. It is not just moths that appear but also parasitic wasps, caddis flies, true flies and beetles, to name but a few groups that are attracted.

I can still remember the first time I set up one. My friend and then colleague Nick Greatorex-Davies, a veteran moth trapper, introduced me to this almost secretive experience. We loaded up the car at dusk with a generator, light trap, collecting containers, torches, coffee and biscuits, and drove into the fading day to the quiet of a local woodland. We unpacked the equipment in complete stillness, the woodlands dampening any sound that we made – silence drifting out of the shadows. The dark silhouettes of the canopy vanished as the night crept in on our yellow circles of torchlight. We sat on the grass, the steam from our coffee rising into the night sky.

Robinson moth trap

That first time the trap we used was the deluxe version known as a Robinson trap. This is a circular container, about the diameter of a dustbin but just eight inches high; it has a mercury vapour streetlamp bulb set in the centre, a transparent plastic lid and a mass of empty egg boxes inside. These provide lots of small spaces for the insects to crawl into once they enter the trap. The mercury vapour lamp produces a cold, intense bluish light as it gives off some UV, as well as light in the visible spectrum.

We sat in this circle of icy incandescence, the darkness pressing in around us while we drank our coffee and waited for our quarry. It was not long after the light had sprung into life that the first moth passed through our bright clearing and, as if caught in some tractor beam, it turned and spiralled in towards the trap. Over that first night, insects came pouring out of the darkness. It is not just the diversity or beauty of these insects that impresses the first-time trapper. The names are the very epitome of British eccentricity, a lexicon of strange elegance. After just one night a host of exotic names were lodged in my memory: the heart and dart, silver Y, large yellow underwing, the sword grass, dusky thorn and burnished brass. Names to conjure with, names that had come fluttering out of the woodland darkness into our oasis of light. Names that are part of the folklore of the nocturnal landscape. Names that are part of the romance of English natural history.

The numbers of insects involved can be overpowering; I have counted over five hundred moths after just one night from a light trap that I ran in my garden in Cornwall. This was not a deluxe version, just a 100 W bulb over a Tupperware cake box filled, of course, with the obligatory egg boxes. Setting my home-made trap became a weekly ritual; plugging it in on Saturday night and emptying it on Sunday morning; identifying and counting the catch before lunch and retaining a specimen of any new species to speed up identification on future occasions. I never ceased to be excited at the prospect of what those dawn walks to empty the trap might have in store.

The local sparrows also came to anticipate this activity and would line up along my guttering at dawn waiting for me to open the trap. I would sit outside my back door with a mug of tea and watch them come. There was always half a dozen already in post, but once I began to walk down the garden, more would drop onto the end of the line, black blobs falling out of a brightening sky. They jostled for position emitting the occasional call. The trap was set on the roof of my garden shed. As I lifted it down, there were always a few insects that made a break for freedom, and as soon as I opened the trap, a few more would slip out into the cold morning air. They would fly low across the garden, a rapid zigzag flight in search of cover. A frantic dash just a few inches off the ground. The squadron of sparrows was poised for action, and at the first sight of an escaping insect one would peel off the roof and swoop after its prospective breakfast. This resulted in a frantic dogfight as insect and bird swerved and spiralled across the vegetable patches towards the hedge. The sparrows came out on top, most times.

The Polytechnic's ecology field trip to the Lake District was based at the wonderful Blencathra centre at Threlkeld, just outside of Keswick. The centre was situated on the lower slopes of the mountain Blencathra and provided amazing views of the Vale of St John and Derwent Water. One of the students' tasks on this trip was to use light traps to estimate the diversity of local moth populations at different altitudes. We used portable traps powered by 12 V car batteries, so for those sites close to a road it was no problem to drop off the gear close by, a short walk to where the traps would be run. One site was a field at the foot of the mountain, another was at the centre, but the unlucky group of students who drew the short straw for the high-altitude trap had to carry their battery and trap 500 metres up the mountain with only sheep paths to navigate by. The catch in these traps was always a revelation to the students when they emptied them just after dawn. The colours and variety of wing patterns never failed to impress. From the soft, downy white ermine that almost invites one to stroke it to the amazing camouflage of the buff tip that looks for all the world like a snapped birch twig. From the black and red of the cinnabar moth to the sleek lines of hawk moths. Breakfast was always accompanied by a line of pots containing the moths that the students were not sure of. Most were no problem to identify, but there were always a couple that required a closer look, and Ken, Mick and I would huddle round the moth field guide trying to extract a name from its pages. Each day the night's catch would be released into bushes around the centre before we took the students out into the field.

The panoramic window in the hostel lounge was wonderful; it provided not only views to impress by day but also a nocturnal spectacle each evening. Once night had fallen, the lights in the lounge attracted many flying insects to this huge pane of glass; moths, beetles and flies would dance on the window a fluttering ballet, dropping in and out of the shadows. The students sat transfixed, but this intense interest in the window display was not restricted to our students. Pipistrelle bats would swoop out of the darkness, hover a centimetre in front of the window while snatching an insect from the glass with their delicate jaws and then peel away into the night. Precision, speed and some dazzling aeronautics, which contrasted dramatically with the uncoordinated flights of the moths mesmerised by our lighted window.

Light traps always impress and nowhere more so than in the tropics. I have wonderful memories of the light trap at Danum Valley Field Centre in Borneo run by the Royal Society. Here there was no pretence at using specially designed equipment to attract insects, just a streetlamp shone onto a white sheet hung by the river. Thousands of insects arrived within the first hour. Local preying mantids took the place of my Cornish sparrows and flew in to take advantage of the banquet; they lined up along the top of the sheet to pick off a meal as a large range of insects was drawn to the light – moths, crickets, assassin bugs, wasps, thousands of small flies. In fact, if you walked between the lamp and the sheet, the cloud of flies would block your nose and send you reeling and coughing into the shadows.

It was at this light trap that I had the unusual experience of being knocked down by an insect. Rhinoceros beetles are substantial insects and, once they gather speed, possess considerable momentum; they are also clumsy fliers and usually stay close to the ground after dark. I was in a half squat with my knees bent trying to photograph the mantids at their dinner when a stray, incoming rhinoceros beetle collided with the back of my knee pushing it out from under me. Beetle and human crashed to the ground, to the great amusement of the assembled company. Humiliation indeed, the entomologist brought down by the object of his desire.

The insects captured in a trap can be literally overwhelming, giving just a hint of the enormous number of species that live in the fields, hedgerows and woods around us. I recall a night that Nick and I spent light trapping in a south Devon wood, intercepting moths as

The light trap at Danum valley, Borneo.

they came to our light and identifying them on the spot. We trapped from 11pm to dawn, which was about 6am, compiling a long list of moth species for the local Wildlife Trust. Then at dawn we decided to open the trap to see what had slipped past us. We took the trap inside our two-man tent to minimise the number of escapees. We zipped up the door of the tent and gently lifted the lid of the trap to peer inside. We could see one or two moths, and the egg boxes appeared quiet, so we thought it was safe to completely remove the lid. Then, just as we had done so, we knocked the base; it was only a gentle tap, but the effect was amazing. The trap erupted, a volcano of moths, beetles, wasps and flies swept past us, filling the air with a blurred mist of whirring wings. I could not see across the tent for insects. There were hundreds, some settling quickly, landing on us producing a coat of many shapes, sizes and colours. Most continued flying, desperately trying to escape the confines of our tent. We looked at each other, clothed in a blanket of insects, and burst out laughing.

Unzipping the tent, we stumbled into the dawn air, a 'smoke' of insects rising from our jackets. Coughing between our continued laughter, we battled to clear the clouds of shed wing scales that had caught in our throats. The liberated moths cascaded into the morning air, a storm of wings vanishing back into the woodland.

Scales and colour

The Lepidoptera, the order of moths and butterflies, is one of the few groups of insects that have scales on their flight wings (some of the weevils have scales on their wing covers, known as elytra). They are thought to be modified hairs that have evolved as a defence against spiders. If a moth gets stuck to a spider's web by a wing covered in scales, it can pull itself free, losing a few scales in the process but living 'to tell the tale' … so to speak. A neat trick, I hear you say, but the spiders are already on the case. The web built by some tropical orb web spiders (family Nephilidae and Araneidae) has a normal top half, but the bottom section is stretched downwards for up to a metre, in some cases forming a ladder-like structure. So, when a moth collides with it, it rolls downward shedding some scales and expecting to fall off the bottom, but the web keeps on going until the moth has lost all its scales, thereby ending up as a victim for the spider. That's the evolutionary arms race for you.

Scales give moths and butterflies their brilliant colours; thousands of tiny dots of colour (like the pixels on your digital camera screen) make up the patterns that we know and love. These patterns are used to camouflage the insect, act as warnings or send signals to potential mates. Butterflies often possess dull undersides and brilliant upper surfaces, so they can hide with their wings up and send coloured flashes to members of the opposite sex when they open and close their wings. Other patterns, such as large eyespots, are defensive; a quick flash of these eyes may persuade a predator to back off.

Light trapping can also take you to places you would not normally go. I was invited to join Adrian Spalding on a hunt for a rare moth, the orange underwing, which feeds on oak leaves and is therefore found in oak woodlands. Adrian, an experienced moth trapper, was the director of the Environmental Records Centre for Cornwall and the Isles of Scilly (ERCCIS). He had been looking for this moth in Cornwall but failed to find it, so he wanted to see if it could be found in the tiny fragment of ancient oak woodland on High Dartmoor known as Wistman's Wood. This isolation has made it a likely location for this rare moth. The wood is truly fantastic, with its twisted trunks and boughs climbing from between great, moss-capped granite boulders. The trees themselves drip with pendulous beards of lichens and are decked with an assortment of ferns. The effect is Tolkienesque, a magical forest that triggers the imagination. Its remote location and mysterious atmosphere have made it famous as a site with an occult connection, and locals tell many tales of mysterious goings-on.

We arranged to meet on an August evening in Princetown just before dusk. After loading the gear into Adrian's Land Rover, we drove across the boulder-strewn moor to the wood, weaving an erratic course between the granite rocks and the occasional sheep. The light trap was quickly set up and we waited for our elusive quarry. It may have been August, but by midnight the temperature had plummeted and we clutched our coffee flasks for warmth. At one o'clock some friends joined us looming out of the night bearing fresh supplies of coffee and biscuits. Still no sign of our elusive moth. Plenty of common species, but the prize remained at large.

August was also a good time to see shooting stars, and by walking a short distance from the trap, the wide empty Dartmoor sky became a black mantle scattered with luminous dust across which bright lines were drawn again and again. An ephemeral script etched by fragments from beyond our world. A script occasionally punctuated by the laborious passage of a satellite. We stayed until three in the morning, and while we recorded a large number of moths, the orange underwing failed to drop by. Driving to the wood had been fun, driving back using just the headlights of the Land Rover became a test of nerve and stamina as we negotiated the labyrinth of rocks and gullies that now stood between us and the road. The simple drive of a few hours ago became a slow progression of backtracks and cautious advances. We eventually got out and guided the vehicle back to the road, torches and flailing arms paving the way. The cold seemed to have seeped into my very bones as I drove home; the heater was on full blast, and I was eagerly anticipating a warm duvet and sleep only to find that the moorland cattle and sheep felt the same way and had gathered on the road where the tarmac was still warm from the day. The journey slowed to a crawl as I wove between the dozing livestock that were now more than comfortable on the winding road. Flashing my lights to clear a path, I drove a laborious slalom around reluctant cattle and sheep as I crept homeward.

I should say at this point that light traps can have a downside as they can be misinterpreted. On one occasion in Cornwall I was light trapping with Nick again. We were setting up the trap at the edge of a wood, not far from the old farmhouse I was renting in the Seaton valley. This was a magical house with an enormous dining room and a huge open fireplace in which we burnt driftwood collected from the local beach. The owners had given us *carte blanche* to roam their land and the go-ahead to collect insects. Nick and I found ourselves struggling down overgrown paths with the trap and generator, then setting up on the valley side as dusk fell, the fading light retreating gracefully down the valley to the grey beach at Seaton. We sat drinking coffee as darkness stole out of the woods and swept away the landscape leaving us isolated in our pool of torchlight.

Time to fire up the generator – a few pulls on the rip cord and it purred into life. The mercury vapour bulb in the Robinson trap flickered and a dull incandescence appeared, slowly brightening into a brilliant blue sun that blazed from our trap. It was too bright to look at and can damage your eyes if you do so for long periods. Our cocoon of light stabilised and soon moths, beetles and other insects were flying into the trap. As the night progressed, we walked around the trap, net in hand, catching, recording and storing specimens that warranted further investigation. Nick showed me several moths I had not seen before, and we quickly built up a list of moths for this area. Time slipped by and at midnight we stopped for coffee. We sat outside the circle of light looking down into the valley. House lights were still burning across the valley and the occasional car headlamps traced the road that threaded between them. A sudden burst of activity caught our attention as the blue flashing lights of two police cars drew up outside of one of the houses. We looked at each other: this was bad news for someone, we thought. A few moments later both cars moved off, the valley was still again, so maybe the incident was not that bad after all. The house lights still burned but the road was quiet.

Back to the catch. The incoming stream of insects had fallen to a trickle of species we had already recorded, so we adjourned to a bank just outside of the pool of light and waited. There was a sound behind us. I turned expecting to see a badger or fox, but instead two policemen were walking out of the darkness. We stood up, somewhat puzzled, and duly offered a greeting while they looked us over. One turned and spoke quietly into the darkness. My landlord and a group of farm workers stepped forward, shotguns and stick laid across their arms. I sensed that all was not well! "What the bloody hell is going on?" my landlord asked in an aggressive manner. Apparently, the people in the house across the valley had reported a witches coven in full session, so the local community had turned out in force to deal with it. We managed to calm everyone down and avoid the ducking stool, and to my surprise, I continued to live at the farmhouse for another year.

Chapter 2

Aphrodite rising – dragons, damsels and river flies

As a small boy, I had been both fascinated and terrified in turn by the large dragonflies that skimmed the lily-strewn surface of the pond in the centre of the small village of Chiddingfold in Surrey, where I grew up. To an eight-year-old boy they were huge, the largest insects I had ever seen. They cruised the reeds at the rear of the pond, then made forays, flying low and fast across the open water to check out the children on the path who were hunting for tadpoles. Everyone knew that they were stingers and that almost certain death awaited anyone who was unfortunate enough to be attacked. Plus, there were rumours, which came from the grandmother of a friend's friend, that if you fell asleep at the water's edge, they would sew your eyelids together (the devil's darning needles). The result was blind panic each time the dragons came to call: flailing nets and arms, spilt jam jars with tadpoles struggling to regain the water, boys leaping in all directions generating a cacophony of shouts and screams. Stealing tadpoles from beneath the noses of the devil's darning needles became a challenge, a test of courage and maybe a rite of passage. Older boys always knew even older boys who had had near-death experiences during encounters with these deadly flying dragons. But despite these fearsome tales, I would often risk life and eyelids to watch these amazing insects. Lying unseen in the long grass at the water's edge, I could see them hunt over the pond, hovering amongst the bull rushes in between high-speed dashes across the open water in pursuit of prey. These scary tales are, of course, untrue, just stories made up or passed on to scare younger children, but this mixture of fascination and respect has never left me. Even now I am awed by their speed, elegance and mastery of the air; hours can drift by on summer afternoons as I sit and watch my childhood nemesis with wonder and admiration.

Dragon- and damselflies (order Odonata), colloquially called dragons and damsels, also have wonderful common names – hawkers, darters, skimmers and demoiselles – names that capture their speed and elegance, names with a certain medieval romance, bold hunters, high speed aeronauts and beautiful dancers. When I first arrived in the South West, I was captivated by the bright, metallic blue of the beautiful demoiselle damselflies that would flitter among the granite boulders that were strewn across Dartmoor's streams, which tumbled chaotically across the landscape. Here was a truly exotic insect, far more flamboyant than dragons of my childhood; flashes of intense iridescence that would dance in the afternoon sunshine, a shimmering ballet performed against the sun-flecked ripples on the surface of moorland streams. The beautiful demoiselle is the common species on the edge of Dartmoor. Its Latin name is *Calopteryx*, a name to conjure with, a name as exotic as its wings are intense. When I was younger, I had sometimes thought that if I had a daughter this would be a great name to give her, but I also imagine she might never forgive me.

We would often walk along the banks of moorland streams to picnic with friends, catching the summer rays with the children playing in the water, undeterred by the almost arctic temperatures. Whilst the children warmed up on the bank, we would watch the demoiselles courting amongst the hot granite stones, the brightly coloured wings of the males contrasting with the transparent wings of the females, which carried only a thin band of blue. Resident males would fight off rivals to keep control of a good stretch of stream – spiralling helicopters of metallic blue as they attempted to drive invaders onto the surface of the water. Pairs of demoiselles would fly linked in a wheel, males clasping females behind the head while the females brought their genital openings up to meet the males', copulating as they flew across the males' territory. Copulation complete, the males held on to their females, continuing to grasp them behind the head while they searched for an oviposition site – attentive to the last. Such casual observations on a summer's day can be misinterpreted through our human perspective, the damselflies' view is very different. The males dare not let their partners go as other males would dive to mate with any available female. Male damsels and dragons give no quarter when it comes to passing on their genes to the next generation. Their genitalia are equipped with either tampers or scrapers to deal with sperm deposited by previous males. This is removed and discarded or pushed to the rear of the oviduct and the new male's sperm deposited in front. Last in, first out.

Despite their abundance, we still have much to learn about our dragon- and damselflies. Back in the 1980s, Mick Uttley, a dragon chaser of old, supervised a PhD student looking at the ecology of damselflies of the genus *Calopteryx*. Here was the perfect research project. Field work is wonderful but can be less than ideal if it demands that you go out in all weathers. However, as the adult *Calopteryx* are only active on sunny days in the spring and summer, the student who took the post could only conduct their research in fine weather, a point that did not escape the other postgrads who had to go into the field, rain or shine. The student who took up this challenge was Claire, and one of her investigations was to estimate the size of the population of damselflies along particular stretches of the leat that runs across the moor above Burrator Reservoir, a workplace to die for. The leat is a man-made channel that diverted water from the high moor to the old towns of Plymouth and Devonport where it was the town's main water supply. It meanders across the landscape, hugging the contours, descending very slowly from moor to the towns. It has been redundant since the construction of the reservoir but still gathers water which now drains into the artificial lake. It is a water course ideal for demoiselles as it is overhung with vegetation and its waters travel at just the right sort of speed. From its banks the moorland falls away, descending to the forest-fringed lake of Burrator. On a good day the lake reflects the perfect blue of a summer sky, framed by the granite massif of Sheeps Tor, Sharpe Tor and, of course, the dam, an artificial tor that holds the lake in place.

As an outdoor laboratory, it has to be one of the best locations I have ever worked in. Claire would undertake a regular count, and in order to make sure that the same damselfly was not counted twice, all damselflies were caught and a number painted on their wings. Counts then became simple and the damselflies' movements along the streamside could also easily be mapped. As many hands make light work, I would often assist with the counting. Arriving as the day warmed up, we would be on site as the damsels stirred into life, the dew fading from the vegetation and the chill lifting from the air as we started the count. It seemed slightly surreal to be hunting for numbered damselflies, and many a passer-by was heard to comment on how remarkable it was that the markings on the damsels' wings looked so much like numbers. Insects by numbers, a new natural history made easy.

In the early 1990s, Mick Uttley and Ken Thompson ran an ecology field course in the Lake District. I was invited along to help with logistics and some insect ID. With Mick involved, it was no surprise that dragons and damsels were on the list of organisms to be investigated. One of the exercises was to estimate their populations around a lake or tarn, just as Claire had done on the slopes above Burrator. Sometimes we used the south end of Derwent Water, at other times it was Tewit Tarn, a small water body nestling in the lower slopes of High Rig, in the Vale of St John. Students were armed with butterfly nets and fine felt markers with which to mark the wings of the insects they caught. We would spend two hours in the morning catching and marking damsels and dragons, have a leisurely lunch, then see how many marked individuals could be caught in the two hours after lunch. Simple really; what could possibly go wrong?

Groups of students patrolled their given stretch of shoreline. Eyes peeled, nets at the ready, taut, alert, coiled springs ready to leap into action at the first whir of a damsel's wings. A shout, their quarry had been sighted, students moved cautiously, closing in on their prey. The stalkers converged, a lightning strike with the net, then pandemonium broke out, the air was full of whirling nets and arms. Bodies collide, shouts, curses, then laughter. Those still standing surveyed the empty nets and their muddy colleagues. This was not as easy as it first appeared. The lazily drifting damselflies had proven to be airily nimble and no easy prey. How would the dragonflies fare?

A military-like operation ensued with students lining up along a known dragonfly flight path which followed a drainage ditch. They waited for the return flight, poised and tense. Almost unnoticed, it was upon the end of the line. A rapid Mexican wave of nets curved across the morning sky. Excess energy propelled some net wielders too far and they toppled into the water. Those still standing collapsed with laughter and the dragons slipped past uncaught. And so the morning went by, with fun, laughter and dragonflies and with thousands of kilojoules of energy expended. The results were often uncertain but always made the point that sampling some invertebrate populations was fraught with problems.

When I lived at Horrabridge in the Walkham Valley on the edge of Dartmoor, my sons and I spent one spring building a pond in the garden. This began as an empty water container, yet within twenty-four hours of filling it pond skaters had arrived. I never looked back, and after the second year one of my great joys was to sit by the pond and eat breakfast while watching dragonflies emerge from their nymphal skins. Climbing out of the water, the larvae would drag themselves up the vegetation at the pond's edge. Once clear, each larval skin would crack open down its back and the adult would haul itself out into the slowly warming day. An amazing transformation would be enacted right before my eyes.

The awkward gothic monster from the depths of the pond would emerge from its discarded skin and inflate its wings. It would then sit patiently waiting for these to harden, trusting to luck and destiny that it did not become an avian meal in these defenseless moments. It had finally emerged to become the sleek colourful aeronaut that we know and love. Once our pond had placed our garden on the dragonfly map, various species of hawker would come calling, checking the pond and whatever else was of interest.

On several occasions I was sitting on my patio with friends when the dragonfly patrol dropped by, hovering just a foot in front of each of our faces in turn, moving up and down as it adjusted its flight. I am sure I saw them tilting their heads to one side, looking almost inquisitive. They would move around the assembled guests, checking out each one carefully, then depart as suddenly as they arrived.

At Horrabridge I also encountered another aquatic emergence, but on a much grander scale. I was leaving the village pub, The Leaping Salmon, having met with a local author to discuss the science embedded in his latest science-fiction story. The pub is across the road from the River Walkham, which runs from the high moor down through the village then joins the River Tavey a few miles downstream.

I walked along the road that runs next to the river, the stands of hemlock water-dropwort forming a dense band of vegetation at the river's edge. The creamy flowers were just beginning to appear as a pale haze crowning the intense green of the leaves. My attention drifted across the river when I thought smoke was blowing across it from gardens on the other side. I quickly realised the smoke was rising from the river itself. Black wisps of ethereal material were spiralling up from the water only to fall back then rise again. The phantom fires that fed them were unseen, but the smoke slowly filled the air above the river. I moved to a small footbridge that crossed the river and discovered thousands of small black mayflies alighting on the handrails. The synchronised emergence of the local mayflies was occurring right there in front of me. This is one of nature's spectacles: entire water bodies give up their adult mayflies at the same time – millions of mayflies leave the river where they have fed and grown over the past year and entered their brief adult phase. Having spent the last year feeding, they leave the comfort of the water for a frantic last night of sex.

The young aquatic manifestations of their lifecycle, known as nymphs, float to the surface and then crack open their nymphal skins. Climbing out onto their old skin, which now acts as a raft, they inflate their wings, ready for the wildest night of their lives. This one-night-stand is a race against time; they do not feed, relying solely on reserves within their bodies to keep them going. Males try to mate with as many females as they can and in so doing rapidly consume their reserves. Exhausted, the burnt-out males fall back onto the surface of the water where local trout are waiting. Mated females hunt for an ideal place to lay their eggs, flying upstream as they do so. The clock is still ticking. By morning they will also have achieved their goal, having scattered their eggs back into the river; after that they are exhausted and dying. Fish are in heaven and gorge themselves on the mass of fast food.

To a naturalist this mass emergence is a stunning sight; to a fly fisherman it is a hopeless time, known as duffers' week. It is a time when fish are so full of mayflies that they ignore even the most attractive of lures and hand-crafted flies. Fishermen can leave their rods at home when the mayflies emerge. I had seen this sort of mass emergence once before on the banks of the River Thames. The trees along the river also seemed to be on fire. Curious tendrils of smoke rose from the canopy of each tree, giving it a wavering, pointed hat – strange mobile shapes created by some ethereal milliner. I stopped the car to investigate further. Walking around the trees, I found tiny flies drifting down from the dark mass above. These were a cloud of chironomid flies, tiny non-biting midges that had emerged from the river and gathered around the treetops. Males lock onto a significant point in the local landscape and aggregate, releasing pheromones that attract females. The ladies drop by and are immediately pounced upon by the males. As I drove on my way, the twisting pillars of dipteran mist were weaving dark silhouettes against the fading light.

Even in the apparent calm and tranquility of a Dartmoor stream, dangers can lurk undetected, risks waiting to be awoken by a chance encounter. So, it was one sunny afternoon when I had taken my sons, Steve and Dave, to the River Plym close to its source at Cadover Bridge on Dartmoor where we were hunting for water monsters and small fish. We had been in the stream for over an hour, the cold water chattering over our sodden trainers, a few small fish and an array of stoneflies and caddis in our trays, when I spotted something interesting in the water. Something shiny, with a square, crystalline look, something that said 'Here is a prize if you reach out and take me.' At this time my eldest son Steve was in his rock- and fossil-hunting phase, so I had visions of presenting him with a large piece of iron pyrite (fool's gold) to add to his collection. I leant down and grabbed the prize, hauling it through the rippling surface of the stream. Reflections of the sky ran between my fingers as the crystals became more regular, a pattern of neat squares arranged around a small, blunt rugby ball. While the outer surface was shiny, the grooves were dark with rust and corrosion. I stared, unable to make sense of this strange object for a second or two, then realisation dawned and my buttocks clenched, my heart rate accelerated. I froze, not daring to move, and turning my hand slowly; I examined the hand grenade I had picked up, looking for a pin. No sign of that or the firing arm. I dare not move; I had no idea about weapons. Was this safe? It had obviously been in the water for some time, or was this a catastrophe waiting to happen?

"Hi Dad, what have you got there? Hey, that looks like a hand grenade. That's cool! Can I have it?" said Dave, jumping in the stream and reaching to take it from me.

My reply was sharp and hostile. He jumped out of the stream and hid behind a bank. I hoped it was the grenade he was hiding from and not me. Gingerly I walked to the other side and placed it carefully in a hollow. I then reversed back across the stream, keeping an eye on it for some unaccountable reason, I grabbed Dave and retired to a safe distance to consider the next move. Steve was mortified and wanted to take a look for himself. No one was going to go near it until someone with a military background tells me it's safe. I decided to call the police, and as it was before the mobile-phone era, I drove to the nearest phone box and called the police station in Tavistock:

"Hi, I would like to report that I have found a hand grenade in a stream close to a public picnic spot."

"Are you sure, Sir? It's probably a lump of fool's gold, this has square bits on it. Looks very similar, lots of it turns up around here." (Now, I don't blame the policemen; the chances of a hand grenade turning up on his shift on a Sunday afternoon was probably very remote indeed.)

"No, it is definitely a hand grenade. It's made of metal, no mistaking the shape and pattern. Can you arrange to have it removed?"

"Sir, I'm sure you haven't, but I will come and take a look. I still think it's a strange boulder you've found." Half an hour later a slightly irritated policeman arrived on a wild goose chase.

"Good afternoon, Sir! Where is this object you think is a hand grenade?"

I led him over to the stream bank. Steve and Dave hung behind, eager to see what was going on but not wanting to fall foul of an irritated policeman. We jumped across the stream to where I indicated the object in the hollow. He peered in, crouched down and reached out with a hand to touch it, then stopped. He stood up slowly and stepped back smartly.

"Bloody hell, it is a hand grenade! Get back, get those kids out of here, clear the area!"

We were hustled back to the car park and radio messages were exchanged in a furious fashion. The bomb squad had been summoned to dispose of it. Steve and Dave were most impressed and wanted to stay for the explosion. We waited for an hour, but nothing happened; we left, slightly disappointed, but with a sense of adventure and adrenaline still trickling through our veins.

Chapter 3

Spiders

Spiders have often been seen as possessing a mysterious or even magical quality, and in the past this was quite literally the case. Back in medieval Europe, the garden cross spider was a talisman against ill fortune and evil spirits which was often worn imprisoned in a pendant hung around one's neck. In the 17th century the good Dr Moffat (whose daughter features in the famous nursery rhyme) regularly prescribed a spider smeared on a piece of bread as a cure for a range of ills, from gout to the ague, whereas in the 21st century it seems curious that the appearance of a Christian symbol on an otherwise unremarkable creature can bestow it with such remarkable powers.

The wasp spider *Argiope bruennichi*

Today spider webs are well known for their symmetrical elegance and beauty, but a closer examination of the spiders themselves will reveal that they also possess an unexpected and discrete charm, a strange beauty of form plus some fascinatingly bizarre behaviours and lifestyles. Here is a group of invertebrates that instantly polarises opinions; mention spiders in any conversation and a schism appears. On one side they are fascinating, beautiful, industrious objects of wonder, on the other they are objects of terror and revulsion, creatures from the realms of our worst nightmares. Few people are ambivalent towards spiders; they are loved or hated, and there appears to be no fence on which to sit when it comes to our attitude to them.

A good test of public opinion is the way that things are portrayed in the world of advertising and literature. Where spiders are concerned there is a distinct hint of schizophrenia. Back in the 70s, the National Savings and Investments Bank used a cartoon version of a money spider in its advertisements as an icon of hard work and persistence, no doubt inspired by the observations of Robert the Bruce as he hid in a cave to escape the enemy. Here spiders are portrayed in a positive light. Then in the 90s, they used a range of spiders, scorpions and insects in a campaign to highlight the lack of financial surprises associated with their accounts when compared to other banks. The advertisement presented the competition as a gallery of spiders and insects asking which has the most unpleasant sting, spiders this time appearing in a negative role.

Hewlett Packard used an origami spider in a web as an indication of complexity and technical excellence in an advertisement for their printers, showing the industrious and technically brilliant face of spiders. While a few years later an advertisement in an anti-AIDS campaign showed a spider in an intimate embrace with a naked woman, thus linking pleasure and terror in a powerfully disturbing image that linked our fear and revulsion of spiders with that of the deadly AIDS virus.

National Savings and Investments Bank advertisement showing invertebrates in a negative light (5 of the invertebrates have no sting).

In literature, J R Tolkien cast spiders as agents of an evil malaise that was creeping across Middle-earth, invading the Greenwood and transforming it into the sinister Mirkwood in *The Hobbit* and then appearing as a terrible guardian of a secret entrance to Mordor in *The Lord of the Rings*. J K Rowling portrayed them as fearsome outcast monsters in her book *The Chamber of Secrets*, a swarm of voracious exiles befriended only by Hagrid and thus leaving everyone else on the menu. Yet, in *Charlotte's Web* by EB White, the orb web spider Charlotte befriends and saves the life of a young pig destined to be slaughtered by weaving words of praise for the piglet into her web. In Louis de Bernières's collection of stories from the English village called Notwithstanding a spider named George is the gardener's pet and the unseen friend of visitors to the potting shed, all of whom confide in the reclusive George as part of their daily routine. Spiders are seen here as part of the natural background of the garden, but most unusually as an 'agony aunt'. Many children's birthday cards depict a friendly smiling spider, proving that spiders can be considered to be cute by younger members of our society. There is a Polish folk tale that depicts a spider providing baby Jesus with a web to keep him warm in his manger, while African folk tales portray the spider god Anansi as a joker and trickster, always making trouble for the humans that he has dominion over. It seems that our attitudes to spiders are as diverse as the creatures themselves, and diverse they certainly are. The recent count of described species is just over 40,000 globally, with 680 of them resident in the UK, and new species are turning up every year.

The public perception of spiders was viewed as a major problem in the world of television when the BBC produced Sir David Attenborough's series *Life in the Undergrowth*, a series of programmes devoted to the natural history of invertebrates. The BBC was not keen to devote an entire programme to spiders as it felt that viewers would avoid the programme or turn it off, thus dropping their ratings. Sir David fought hard to keep it in and altered the programme theme to that of *The Silk Spinners* thus diffusing the negative perception. The programme started with silk moth larvae and weaver ants before moving into the world of spiders, a gradual lead in to keeping the viewer calm before introducing them to the delights of this fascinating group of invertebrates. Unexpectedly the programme achieved one of the highest viewer ratings ever recorded by the BBC for a natural history programme.

I first became interested in spiders in the early 80s, when as a new member of staff at the then Plymouth Polytechnic we would take students into the field and show them the various invertebrates that we encountered in the woods and moors of the South West. We would reveal the secret life of beetles, flies, millipedes, woodlice and other invertebrates that we encountered, but when we found spiders there was a silence. We had expertise in most of the other groups and, even though spiders were abundant everywhere we went, could offer only rudimentary comments concerning them. It was also well known that spiders were difficult to identify and that few people were rash enough to study them. Here was a challenge, a gap in need of filling, so I set myself the task of doing so.

My official introduction to spiders was on a Field Studies Council course at Box Hill in Surrey, which was run by the late Francis Murphy. Francis was the 'godmother' of British arachnology, an international expert with a passion for encouraging others to become enchanted by her muse and take up the study of spiders. She was a cross between your favourite aunt and a truly terrifying headmistress – kind, patient, encouraging but would tell you firmly where you had gone wrong and sent you back to start again, even if it was your tenth attempt to get it right. Francis was loved by all and is greatly missed.

Life at Box Hill back in the 80s was wonderfully British, a step back in time to a more relaxed age. The Field Studies Council Centre was a large house set in the grounds on the lower slopes of Box Hill. The ground floor had a suite of oak-panelled rooms with large, comfortable armchairs, a wonderful library and a grand staircase that swept out of the entrance hall. Tea and cake were served at 11am and then again at 3pm, dinner was at 8pm; staff and students dined together, turning each meal into another casual tutorial. We lived and breathed spiders for the entire weekend.

Over the course we visited several different local habitats to collect spiders. We generally stayed an hour or so in the field, then it was back to the laboratory with our vials of alcohol-preserved spiders to identify. Here we peered down microscopes looking for characteristic eye patterns, various arrangements of spinnerets, counted foot claws and hunted for the elusive trichobothria. These are extremely long and fine hairs that sense airborne vibrations and air movements. They are arranged along the various sections of a spider's eight legs; each leg section possesses trichobothria of different lengths that respond to different frequencies of vibration. The information from these trichobothria is then integrated to provide the spider with a dynamic view of the local environment. The position of these fine hairs on each section of spider's leg is also crucial in determining which group of spiders one is looking at, especially in case of the tiny money spiders of the family Linyphiidae. In this group the position of trichobothria on the front leg is crucial in determining the genus to which the spider belongs. To be able to locate and measure the trichobothria was an essential skill for anyone who wanted to identify spiders to species level. Thus, we spent hours hunting for this gossamer, thin hair that always eluded the apprentice, a hair so fine it was only seen when one was not looking for it, a touch of Heisenberg's uncertainty principle, a ghost in the arachnological machine that was always just out of reach…until that one time when the merest hint of an image drifted briefly into focus before vanishing. Then frantic re-focusing occurred followed by the exclamation of "Is that it!" Francis would look down your microscope and lay her hand on a shoulder. "Yes, that is it." A line had been crossed; the grand Mistress of spiders had acknowledged your first step on the arachnological highway which now lay open before you.

Few spiders are easily identifiable to species level with the naked eye. Once you have examined a few specimens, some of the more obvious species start to become familiar. But to identify the rest to species level with certainty, one has to examine the genitalia. In spiders these are external, unlike other groups of invertebrates that have internal genitalia, which means the entomologist has to dissect the insect before an identification to species level can be made. Female spiders have their genitalia, known as an epigyne, on the underside of the abdomen, while males sport theirs at the end of their pedipalps, small leg-like appendages at the front of their bodies. The observed genitalia are then compared with drawings of known species in a series of standard texts. This is a tricky process as you are comparing three-dimensional structures with two-dimensional drawings. If your specimen is at the wrong angle, it will look very different from the drawings, so great care was required. This was a nightmare at first, but by the time I left the course I was almost confident.

A little optimistic I fear, as it took me a couple of years to realise my naive arrogance and to then gain enough experience to become truly confident in my identifications. It was also a point of embarrassment explaining to my family on my return home from that weekend course that I had spent much of the time examining spider's genitalia and looking at hundreds of drawings of the same. This preoccupation with sexual organs comes with the territory; knowing your spiders means recognising the structure of their reproductive organs. Over the years I had accrued in my laboratory a shelf full of spider identification guides which deal with spider fauna from across the globe. Many contain photographs or drawings of spiders, but all contain drawings of their genitalia, some are just a collection of such drawings arranged according to their taxonomic groupings. Students working in my laboratory were intrigued by this collection of exotic artworks and would affectionately refer to them as Pete's spider pornography collection.

Spiders are often perceived as mysterious as they appear in unlikely places and walk on almost any surface. They build webs across impossible spaces, even floating on a summer's breeze without the aid of wings. The secret to this almost supernatural ability is silk. Like the American super-hero Peter Parker (Spiderman), silk gives the spiders great power, but unlike Mr Parker, they have absolutely no sense of responsibility; spiders do as they please. Spider silk is one of the most ethereal threads in the natural world. It is everywhere. I am sure that many of you will have caught sight of a half-seen silver thread out of the corner of your eye. A thin strand of gossamer that drifts lazily across the air of early mornings or summer evenings when the sunlight sweeps almost horizontally across the landscape, golden strands dancing slowly on the fading heat of another day. Spiders produce silk everywhere they go, leaving a trail of gossamer in their wake, and as they have been recorded at densities of up to 500 per m^2 in UK meadows, the result is a carpet of silk.

One of the most dramatic demonstrations of this that I had witnessed was in a hilltop forest clearing in northern Borneo. I was there at first light, watching as the surrounding trees drifted out of the darkness revealing their impenetrable shadows, the treetops silhouetted against the ever-increasing brightness of the coming day. Stark shapes against the vivid oranges of the rising sun. The clearing was dark until the sun's almost horizontal light caught the tips of the grass blades, instantly transforming the clearing into a field ablaze with a dazzling blanket of spider silk. A fiery translucence that undulated in the barely perceptible breeze that edged in behind the sun's rays. A spectacle that is as ephemeral as the dawn. As soon as the temperature began to rise, the dew that revealed the webs was driven off and the silken webs faded to invisibility, phantom traceries waiting for the onset of another sunrise.

Of course, you don't have to travel that far to see this amazing spectacle. When I lived on the edge of Dartmoor, I would often walk across Roborough Down early on a spring morning when the dew was heavy on the grass. There the gorse would be laden with

Spider silk at dawn in a forest clearing, Borneo.

diamond-studded webs that sparkled in the early morning light, tiny hammocks strung between the spiny branches or broad silken platforms that vanished into sinister funnels. Still and empty in the quiet of the dawn, the spider's lairs were jewelled lace traceries misted by the onset of the day. Here was a landscape of fabulous riches abandoned by the night but as temporary as my tropical clearing. Once the sun had crept into the moorland sky, the field of jewels evaporated leaving only hints of their former glory.

In a spider's world silk is their Swiss Army knife. It is used for almost everything. It can catch their food, keep their offspring safe and it helps individuals to find a mate or escape from danger. Most spiders have six silk-producing glands at the rear of their abdomen, each one producing silk with a different molecular composition and a distinctive set of properties that makes it ideal for a particular job. Each of these glands is connected to a spinneret, a flexible nozzle that controls the thickness of the thread. Silk is pulled from the tip of the spinneret, thus drawing fresh liquid silk from inside the gland through the narrow nozzle where it is transformed from a fluid to a solid thread. While we do know the physical and chemical structure of spider silk, this mysterious transformation is not yet fully understood.

One gland produces silk that is ideal for the structural threads of webs, tough and inflexible, to keep the web firmly in place, while another produces silk for the spiral capture thread that we are so familiar with in our garden orb webs. This is extremely flexible in order to absorb the impact of flying or jumping insects that crash into the web. Other glands produce the silk that spiders wrap their eggs in to protect them from the elements and potential predators. One kind of silk provides an insulating layer around the eggs, whilst another acts as a tough impenetrable outer layer. This diverse use of silk has been a key factor in making spiders the incredibly successful organisms that they are today.

The trail of silk that spiders leave in their wake acts as a trailing safety line anchored to the surface at regular intervals. These fixing points are another kind of silk, produced for just this purpose. So, if the elements sweep them away, they can simply climb back up their safety line to the last point of attachment and carry on. This silken line is also useful to spiders that hunt by sight or touch when they are looking for a mate. Females simply leave a pheromone coating on their silk and males will then follow this irresistibly perfumed trail.

The reason that silk has so many uses is due to its amazing physical properties. It is stronger than an equivalent diameter thread of steel yet can stretch further than a similar diameter thread of elastic. In the past, many attempts have been made to develop a way to harvest silk from spiders. A tailor in the court of Louis XIV collected webs from the surrounding countryside which he spun into thread from which he wove a pair of trousers for the king. A one-off due to the time and energy involved. Various spider-spinning machines have been constructed to extract silk from live spiders. These pulled silk from the spinnerets and wound it onto a reel, but none could produce enough silk to make it worthwhile. For a project based in Madagascar silk was collected from thousands of individuals of the large golden orb weaver in the genus *Nephila*, a spider which is famous for the intense yellow colour of its silk. Over an eight-year period, this vast collection of silk was spun into thread and then woven into a cloth that was fashioned into a poncho-like garment. This was then embroidered with the most delicate appliqué designs that depicted the spiders that had donated their luminous thread. The result was a garment that had a fabulous intensity, an almost solar radiance. It was duly exhibited at the Victoria and Albert Museum in London for six months in 2012 where it attracted tens of thousands of visitors, and it continues to tour the world's museums today. A Chinese physicist has extracted silk from *Nephila* spiders and twisted it into strings for the violin. These are said to produce a superior tone to existing strings but, due to the lengthy collection process, a production line is unlikely to emerge!

Spider silk is nature's super fibre and has recently become one of the holy grails of the biotechnology industry. The scientist who discovers the secret of how to make synthetic spider silk is destined for fame and fortune. Attempts to do so have ranged from the bizarre to the surreal. British scientists pioneered this research by transferring the genes for silk production from an orb web spider to the bacterium, *Escherichia coli*. Silk could then be extracted from a bacterial broth. This worked, but only small amounts of silk were produced. The Chinese tried transferring these genes into silkworms (the caterpillars of the silk moth, *Bombyx mori*) in the hope that they would spin their cocoon out of spider silk rather than ordinary silk, which could then be harvested in the traditional way. This process produced silk that was a halfway house between the two kinds of silk. Yet, undeterred they are now using new gene-editing technologies and the work continues. American scientists transferred the silk-producing genes from spiders into goats, so that when the goats were milked spider silk emerged in their milk! This was beyond surreal. I did not believe it when I first read the reports. Even Salvador Dali could surely not have dreamt of this. Silk was produced but not in commercial volumes; however, the goats live on and the research continues.

So why is there so much interest in spider silk? The combination of strength and flexibility is highly prized by the military. If a single production process can be varied to produce either very strong or very elastic fibres, a mixture of these can be used to produce more efficient bullet-proof vests, strong but flexible ropes or clothing resistant to wear and tear. There is even a hint from recent research that spider silk possesses some anti-bacterial properties. So, spider silk is destined to be a scientific hot topic for some time to come.

Spiders are common co-inhabitants with us in our houses; these are well-recorded in the literature. Back in the late 80s my colleague Ken Thompson (now famous for his popular science books and gardening columns in *The Telegraph* turned up in the lab one morning with a matchbox. "A little gift for you", he said. Carefully opening it, I was thrilled to find the legendary spitting spider *Scytodes*. This was a spider I had read about but never seen; it is strange looking being straw coloured with black markings and with a large, domed head, known in the trade as a cephalothorax. This enlarged head contains modified poison glands that produce a viscous gum as well as the original venom. *Scytodes* does not spin a web; it walks around the walls of old buildings at night hunting for prey using a special prey-capture technique; it is the 'Sid Vicious' of the spider world that has raised spitting to a deadly art. On locating a potential prey, *Scytodes* creeps slowly forward to get within spitting range. Once it is close enough, it squirts a stream of gum from its fangs and vibrates them as it does so. This has the effect of launching a pair of zigzag streams of gum that act like a net. *Scytodes* is a crack shot and hits the target almost every time. The sticky threads land across the prey, pinning it to the ground. This leaves *Scytodes* time to amble over to it and consume its meal parcel in a leisurely fashion.

Apparently, Ken had found it wandering along his study wall late the previous night. I was amazed that a spider I had thought quite rare could turn up in Plymouth and wondered what else was lurking in the houses of our fair city. My curiosity was piqued, so I placed a request in the Biology Department's newsletter asking my colleagues to bring in any spiders they found wandering inside their houses. To my surprise, I received an avalanche of jam jars containing spiders from the city and surrounding areas. I was amazed by the diversity of spiders found in local houses, especially as the literature implied that there were only a few species that were found in houses. This small survey clearly indicated that other species could be regularly found co-habiting with us.

I was intrigued. Having had a successful response to my local survey, I wondered if this could work on a national level and wrote an article for the newsletter of the British Arachnological Society asking members to send in samples of spiders collected from their houses over a period of a year. I am not sure to this day how the genie escaped from the bottle but escape it did. The media caught a whiff of a whacky story and came calling. While initially I was flattered by the attention of the local media who thought this was a story with a generous helping of academic eccentricity, I was totally overwhelmed when the media storm broke and found myself in a national spotlight. The office phone rang constantly, my colleagues who shared the office with me gave up answering it as it was always for me. Every local radio station in the land seemed to want to interview me, and even the broadsheets sent reporters to talk to me. The *Really Wild Show* came to interview me in my lab, and I even found myself talking to the editor of the *Sunday Sport*. It became scary when chat show hosts asked me to drop by and tell them about spiders. As a man with a stammer, this was my worst nightmare and I declined them all. All that is except for the *Richard & Judy* show who wanted to make a short film and then do a live interview. I decided I would do the film but offer a colleague from the Liverpool area to do the interview. The producer agreed and arrangements were made for me to fly to Liverpool to meet Fred Talbot, the TV weatherman famous for his outrageous jumpers and floating map of the UK from which he delivered his forecasts. Now here was a doubly exciting prospect, being involved with one of the nation's favourite chat shows and also flying for the first time. I had flown in gliders as a teenager but never in a powered aircraft. I think I was more excited at the prospect of the flight than the making of the film.

The day arrived and I drove to Exeter airport at the crack of dawn, checked in to wait to be called. Exeter airport was tiny back then; the departure lounge was more like a dentist's waiting room but without the angst. Half a dozen passengers waited with me. The call came and we walked out onto the tarmac to our Dash 7, which waited patiently, props silent. I climbed aboard and was offered a window seat. I settled in, fastened my seat belt and looked around. The cabin was small, in fact it reminded me of the rear of a large Land Rover. The engines roared into life and, as we taxied down the runway, the whole cabin began vibrating; this was real flying. A brief wait, then full power was applied and we hurtled forwards, the entire cabin shaking and rattling, whilst the engines were attempting to drown my emerging thoughts that it may have been a mistake to step aboard. The ground dropped away and we cruised over the Devon landscape; I recalled from my days in a glider that the noise stops once in the air. I waited for the noise to fade but nothing happened. This was not like a Land Rover, rather it was flying a Land Rover! We rattled and bumped our way across the sky to Manchester airport where I was met by a driver from the TV company and taken to Liverpool. We went straight to the dockside studios where an ebullient Fred Talbot greeted me and whisked me off for coffee and a chat with the producer. The plan was to make the film back at his house, but he had to do the midday forecast first. So, coffee in hand, I stood on the dockside and watched Fred, now resplendent in his latest jumper, leaping across a floating map of the UK as he delivered his forecast for the next few days.

We spent the afternoon filming spiders around Fred's house then flew back to Devon. My colleague James Wright, who lived in Liverpool, had kindly agreed to do the interview, so we waited for the film to be broadcast alongside the live interview. A slot had been scheduled for the following week, however, alas, Prime Minister Margaret Thatcher and her former Defence and Environment Secretary Michael Heseltine went head-to-head over the leadership of the conservative party. As we were but small fry in the media eye, these national events swept us to the back of the queue. Next week was now our slot, but then Iraq invaded Kuwait and spiders again faded from the media's gaze.

But not everyone in the media had forgotten the house spider survey though; Allan Corran chose this moment to weave the house spider survey together with a few other stories in his column in *The Times*. These included a shortage of underpants in Wormwood Scrubs prison and a man jailed for failing to feed his tarantula. All these events came together in a sharp and witty review of the week in which he concluded that he dare not risk sending me any spiders in matchboxes for fear that he might end up in the Scrubs with no underwear! This was one of the most exciting events to come out of this affair, a mention in the column of my satirical hero.

The house spider piece finally went out on the *Richard & Judy* TV show four weeks later. The film was fine, and James handled the interview well. But by then the media had forgotten the quirky spider story. A day is a long time in the media world, but four weeks is an eternity.

Unexpectantly, the media whirlwind generated hundreds of letters of interest. So, over the next year, a trickle of jam jars and plastic containers filled with a year's worth of spiders arrived in my lab. I recall one in particular from a local lady, Sylvia, well known as a life model for local painters who often described herself as a slow-motion streaker. She lived in the cottage at Vixen Tor on Dartmoor. This was a truly remote house that sits at the foot of one of the region's most dramatic tors, famous as the setting for the legend of the evil witch Vixena who would lure travellers to their doom in the surrounding mires. Sylvia's collection of spiders grew in a jam jar that sat on her mantelpiece. She told me that visitors would often enquire about it and express surprise at her explanation that she was taking part in a national survey. Shortly afterwards the word was out that a practicing witch was back in residence at Vixen Tor Cottage.

One year later I had accumulated just over one hundred jars of spiders from across the UK. The results were built into a huge database that showed that a large number of spiders came into people's houses and that rural houses had a greater diversity of spiders than town houses. I recorded a total of 125 species, three of which are only found inside buildings, 26 that were associated with human habitation but were also found outside and 96 that had no associations with buildings at all and were thought to have wandered indoors in error. The results were presented at the annual meeting of the British Arachnological Society (BAS) that took place in Edinburgh that year.

Shortly after I took an interest in spiders, the BAS launched the National Spider Recording Scheme in association with the Biological Records Centre (BRC) at Huntingdon. The BRC is an organisation that documents the fauna and flora of Britain, producing atlases of their distribution, and now spiders had fallen under their spotlight. The person who had persuaded them that spiders were worth looking at was Clifford Smith, a founder member of the BAS and a man with enormous energy. He had already mapped the distribution of the spider species found in Yorkshire and was now 'calling arachnologists to arms' to do the same for the rest of the UK.

The plan was announced in the society's newsletter, forms and maps were made available, and a meeting was called to set up the vast machinery that would collect information on our spiders from across the land. Full of enthusiasm, I offered my services and persuaded the university to fund my attendance of the meeting in Cambridgeshire. This was very exciting as I had never attended any of the society's meetings before and so here was a chance to meet the great and good of the spider world.

The meeting lived up to my expectations. I arrived late for dinner having travelled almost the entire width of the country, but food was kept for me. I ate at a table with a couple of other latecomers and assumed that they were new guys like me as everyone else was socialising in the collection of large, comfortable armchairs in the nearby lounge while others were liberating bottles of beer from the bar. Conversation rolled by and everyone seemed very friendly. Clifford Smith walked over and introduced himself; he too was very friendly and affable. "The meeting starts in half an hour. See you in there. I will leave you in the safe hands of Dr Roberts." I looked at the man opposite, Dr Roberts; the name rang bells. The recently published book *The Spiders of Britain* (which is now the bible for anyone studying British spiders) was written by a Mike Roberts. He looked back at me. I re-wound our conversation in my head to see if I had said anything too crass. He was still talking to me, so I assumed not. I asked him about the book, and he told me how he had fitted writing it around his work as a GP. I was now even more impressed. I had seen an article about him in one of the broadsheets. "Yes," he said, "that was funny; they asked all sorts of questions but were keen to know why I had a WW2 gas mask on the back of my study door." "In case the dog farts," he had replied. I relaxed; anyone who can have fun with the press must be okay.

At the meeting, Clifford outlined the plan for ten years of future field work, followed by a great gathering up of the data and the publication of the *Atlas of British Spiders*. The UK would be divided into counties and people would be appointed to take charge. The person who was then in charge of the South West region of England was Paul Hillyard, curator of arachnids at the Natural History Museum in London. Meanwhile, the rest of the land was to be divided between members of the Society, with a remit to coordinate field work and gather as many records as possible. I offered to organise some collecting in South Devon, while the four Scots present divided the land north of the border between them. I felt that I had come off lightly with just half a county.

Paul Hillyard arrived the next day and we discussed meeting up in Devon. He agreed to come down to a meeting in East Devon if I could organise it. I returned home and contacted all the Devon members of the BAS suggesting a meeting at Aylesbeare Common, an RSPB reserve on the Devon-Dorset border. This meeting went according to plan, and we were duly welcomed by our host, Peter Gotham, the RSPB warden who is a large, genial man with a generous beard, the sign of a true biologist. We collected spiders all morning and then adjourned to Peter's orchard where his wife had prepared an amazing spread of homemade bread, jam and cakes. This was an idyllic moment; lunch in a sun-dappled orchard with a small band of arachnological heroes on the brink of a great biological adventure – an entire county to tame and only ten years in which to archive this. Here was rampant optimism on an unprecedented scale; what could possibly go wrong? Replete with fine food, we returned to the Common and continued to collect spiders. By mid-afternoon, Paul announced his departure back to London. He did not feel he needed to return to Devon and was happy to leave the county in my hands.

Ah, and so it was that I was given Devon to survey for spiders. This was the beginning of Sunday field trips to collect spiders across the county. We initially focused on Devon Wildlife Trust reserves but later expanded to the National Trust and other private landowners as the years drifted by. These Sunday field days were fabulous days out. Our small band of spider hunters would wax and wane; we had an amazing variety of people, everyone from computer programmers, ex-naval missile experts, yacht salesmen, university students, medical consultants and even the head of the land registry passed through our ranks. We were enthusiastic, relaxed and possibly eccentric but derived immense pleasure from those days in the field hunting for spiders.

While our field days occurred over the spring, summer and autumn, the subsequent lab sessions began as follow-ups, but as the backlog grew, they became a fixture in our weekly calendars. Wednesday evenings in the University laboratories identifying the catch was a regular event, fifty-two weeks a year. Martin George, the legendary Ivor Kenny, myself and anyone else who was free would assemble in the lab at 6pm, work through as many spiders as we could until 9pm and then adjourn to a local pub for refreshment. New species were added to a rapidly growing reference collection of specimens that we kept in order to confirm our identifications. This collection now resides in the archive of Plymouth City Museum, known as The Box

It was just a year before the Devon survey was launched that a retired toolmaker had wandered into my lab asking for information about spiders. He had tried the city library but failed to find anything useful. He understood from a friend of his ex-wife that I could possibly help him, and so started a friendship between Ivor Kenny and myself that lasted until his death in 2012. Ivor was fascinated by the detailed observations required to identify spiders. As a toolmaker, detail had been everything, so arachnology was a natural progression. He had also been an artist, jazz musician, milkman and publican before his final career as an arachnologist.

Conversations of a Wednesday evening would range over the biology of spiders, of course, but then on to the wider world: the Newport jazz festivals and jazz in general, Ivor's life in the early tank regiment and the maths behind the golden section and his life as a painter. Ivor attended almost every field trip we made over the fifteen years that we gathered information for the atlas. He was a man who did not compromise and, while the rest of the crew opened a thermos of coffee to wash down their lunch, Ivor always cracked a can of Bass. He found some field sites more challenging than others and complained bitterly about the tall tussocks on the culm grassland sites. Years later he presented me with a copy of a paper on money spiders in which he had written "To my friend Peter. I forgive you everything, even the culm grasslands."

As the deadline for data collation approached, we became aware that we had a large backlog of samples to clear. So, Ivor and I decided to go into the lab on Sundays to attempt to redress this. This worked well over the autumn, but when we entered the cooler months, it became more challenging. By the time January arrived, the university's economy drive hit home as they turned off the heating at the weekend. We ended up working in our hats, coats and gloves. Our breath would form clouds which would condense and fog the eye pieces of our microscopes. These freezer-like conditions eventually drove us from our academic pursuits to engage in deep intellectual exchanges in the nearby pub.

Ivor's attention to detail, combined with his training as an artist, had also meant that he was soon drawing the spiders that we found.

Once the first of these was published in the Arachnological Society Newsletter; he received regular requests from the editor for drawings to accompany articles that had been submitted. Memories of his talent, humour and enthusiasm are forever linked to the many evenings we spent identifying spiders, talking jazz and drinking beer.

By 2001 enough data had been collected from across the UK to bring it together and produce a set of maps. These were then edited, and each of the recorders were allocated a group of species to write the accompanying text that would outline the species biology and conservation status.

The Provisional Atlas of British Spiders was finally published in two volumes in 2002 and remains the defining document regarding the distribution and natural history of our spider fauna. Fifteen years of hard work, good times and new friendships had produced a milestone in our understanding of UK spiders.

A spin-off from the spider atlas came from our visit to the quarries at Buckfastleigh. Whilst surveying the area for the atlas, we were invited to take a look at the spiders living inside the locked and gated caves that opened into the limestone cliffs. We were given a tour by Wilf Joint himself, the caver who had opened up the caves and discovered a treasure trove of ancient bones from animals dating back

The cave spider *Meta menardi*.

to before the last ice age. The caves were large with plenty of room to stand, and Wilf pointed out the large cave spider *Meta menardi* that hung from the upper regions of the cave.

Here was an impressive spider with a leg span of around 5 cm. I became fascinated by these denizens of darkness, and on my return to the Polytechnic, I delved into the literature and was surprised to learn that almost nothing was known of their biology. This seemed strange for such a large and impressive creature, so I set about organising a project student to work with me and illuminate the intellectual darkness that surrounded the cave spider. Philip McKenzie offered his services, and we were soon regular visitors to the caves. There were so many questions: What did they eat? How did they catch their prey? How did they find new caves? What happened to the young spiderlings?

Philip decided to look at how spiders of different ages distribute themselves around the cave. This seems quite simple: just measure the distance from the entrance and record the size and sex of the spider. He quickly discovered that caves possess complex surfaces with many nooks and crannies, some of which were impossible to get to, so mapping the position of a given spider was not as simple as we first thought. To make it easier, we changed location and began work in an abandoned mine near Mary Tavy on the western edge of Dartmoor. This was a very linear entrance tunnel carved into the granite and hence was very much simpler to map. Philip was able to determine the whereabouts of adult males, females and immatures over the late autumn and early spring. He discovered that the females move away from the entrance in the winter, possibly to avoid lower external temperatures, but then move back in the spring as external temperatures rise. While the external temperature would vary wildly over the winter, those inside the mine remain stable at between ten and twelve degrees centigrade.

The results were published in the bulletin of the Pengelly Caving Trust later that year. Phillip graduated and moved on, but I continued monitoring the old mine and did so for the next three years. Once a week I would visit the mine before going in to work to map the spiders' positions and steal any prey I discovered in the webs, preserving it in 70 per cent ethanol for later identification back in the lab. The prey items varied widely. I identified a range of beetles, millipedes and flies, but one prey item kept eluding me; a squidgy mass of slime that hung as a partly digested droplet suspended by a single thread from the centre of the web. These turned up on an almost weekly basis, and it was some time before I found an undigested version – which turned out to be a slug. Cave spiders, it seems, have a distinct fondness for the gardener's bane. Collating all the data, it became clear that *Meta menardii* preyed on invertebrates that crawled over the cave or mine wall rather than those that flew in. The data I had collected on the distribution of the various life stages of this spider in relation to time of year also allowed me to construct the life cycle of this spider; the project was a great success.

While engaged in the process of collecting data for the national spider atlas, I had approached the National Trust for permission to sample on some of their sites. As a result, they contacted me the following year to invite my team to visit a few more of their reserves but now with travel expenses, an offer that I could not turn down. This went well, and I sent in regular reports of the spiders that we found. A few years later they approached me again asking if I would be willing to look for a couple of extremely rare spiders that had been recorded from their land on the Lizard in Cornwall. These were the ground spider *Gnaphosa occidentalis*, which had not been seen since the 1930s, and the other, the small sac spider, *Clubiona genevensis*, had not been seen for at least twenty years. Here was a wonderful opportunity to walk some of the South West's most dramatic coastline and hunt for spiders. A quest for arachnological unicorns in the wildest landscapes of the region; the romantic in me leapt at the prospect.

The Lizard peninsular is one of the most diverse and beautiful tracts of the Cornish coast. It is dotted with small coves, ragged cliffs and hidden harbours. It boasts the most southerly point in the UK whilst its neighbour, the Lands' End peninsular, boasts the most westerly. On the Lizard the flat, wind-blasted heaths of Predannack and Goonhilly Downs run down to the ragged coves of Kynance and Lizard Point, but there are also the tiny, sheltered harbours of Mullion and Cadgwith, plus the enchanted wooded valley of Poltesco. Not only are they ruggedly beautiful landscapes but the very names have a wild romance all of their own.

The National Trust owns almost two-thirds of the coast on the Lizard, so there was a lot of territory to cover. As there are not many access points, driving from cove to cove would be tedious and involve retracing one's steps so often it seemed unfeasible. Hiking around the coast appeared to be the most practical option and it meant we could sample more frequently. For these visits I would employ a helper; students from the University were obvious candidates and both my sons were also keen to take part. Over the four years that I visited the Lizard, my sons Dave and Steve came along to assist, while two students from the University, Barry Green and Gareth Prowse, also did a stint at carrying the gear. The landlady of the diving lodge where we made our base would drop us on the coastal path in the morning and then wait for a call to collect us from a point further round the coast at the end of the day. The task was simple: start at Gunwalloe cove and walk as far as the sampling allowed that day, terrain and spiders permitting. We would stop every fifty metres or so and hunt through the cliff-top habitat for spiders. We also stopped at other places that looked promising. We would turn over stones, use nets to sample the taller vegetation and any trees or bushes that we encountered, and used a suction sampler to collect any ground-dwelling spiders. The suction sampler is a modified petrol-driven leaf blower that has a fine net installed in the intake tube thereby turning it into an invertebrate vacuum cleaner which is known in the scientific literature as the G-Vac. Using this device, we could suck up any invertebrates that were hiding in the base of the vegetation. Once a sample was taken, the net in the intake was emptied into a white tray and the catch examined; any spiders were collected and preserved for identification back in the lab.

We set off armed with our collecting gear, specimen tubes, alcohol to preserve our catch and notebooks to record everything of interest, plus copious supplies of water and, of course, pasties from the legendary Ann's Pasties shop in Lizard village (field work in this part of Cornwall is unthinkable without these vital supplies). Carrying the two-stroke driven G-Vacs seemed almost effortless at the start of each day, but as the day progressed, the weight gradually increased, reaching at least ten times its original weight by the time dusk was approaching. The G-Vac comprises an engine fixed to a long, black plastic tube which makes it unevenly balanced; the best way to carry it is on your shoulder. So, as we traveled the coastal path, we provided other walkers with an unexpected spectacle as we emerged from the heat haze on the heath in a somewhat Rambo-esque fashion, carrying what appeared to be a piece of serious military hardware on one shoulder. Sometimes questions were asked – a polite enquiry that may turn to real curiosity or more often a swift departure, others passed hastily by. In either case it was always accompanied by a polite greeting, even if it was followed by a speedy farewell. It appears that the British acceptance and respect for eccentricity is still alive and well.

These walks were fabulous. Each day was a hike through one of the most dramatic landscapes I know. The tiny church perched upon the cliff-top dunes at Gunwalloe overlooking the small perfectly formed bay was a wonderful starting point for our annual journey. A journey across a mysterious landscape, from the monument that commemorates Marconi's first transatlantic radio transmission to the abandoned airfield at Predannack, still haunted by the rusting remains of the aircraft that once cruised the skies above the wild heaths. Down into Kynance Cove where the waves crashed onto the naked serpentine rocks, and then on to Lizard village with its lifeboat poised on its cliff-top station just waiting for the boatswain's call. The adjacent cliffs were ablaze with colour from the invading

Hottentot fig that cascade down the rock face towards the sea. On, around the coast to the rock arch known as the Devil's Frying Pan, the sea roared in and out of the collapsed cave that is now a small cove behind it. Then descending into Cadgwith Cove, where fishing boats are still hauled onto the beach in front of the Old Cellars pub. On to Poltesco, where a magical stream winds down the tiny valley to the sea via a fairy wood and abandoned mine buildings, a journey halted in its tracks by a pebble beach that had been built into a wall by an impetuous sea, a wall that is higher than a man. Then around to the final dramatic sweep of Chynhalls Cliff and down into Coverack where a welcome halt awaited us with fine coffee and cakes.

The four visits over three years provided an excellent inventory of the Lizard's coastal spiders, recording 105 species of spider, seven of which were of conservation interest. The rare sac spider *Clubiona genevensis*, which had only been found at two sites previously and had been unrecorded for twenty years, was relocated at both of the previous sites plus two others. Yet, the very rare *Gnaphosa occidentalis* was never located on these quests and remained lost for another ten years until John Walters and I eventually discovered it on a cliff-top path at Penhale in Cornwall.

When the Eden Project opened, I can still recall the air of excitement and astonishment that such an undertaking would happen almost on our doorstep. During those early days, Tim Smit came to the university and delivered a lecture in which he outlined his hopes, aspirations and ambitions for the project. He was, of course, extremely eloquent and described an exciting future with many possibilities for research and collaboration. This sounded exciting; however, I began to wonder with all those tropical plants being brought in from around the world what kinds of invertebrates might tag along. As the grand opening approached, several of our graduates gained positions on the project; one of them, Cathy Trodd, relayed to me the news that the science team did have an interest in the invertebrates that might become established in the domes. We had a meeting with them and were granted permission to put

The Eden Project domes.

down pitfall traps in the tropical dome. These stayed down for a couple of weeks after which we collected them and identified the catch. A range of invertebrates were identified and a short report passed to Eden's science team. These results were also published in the *Bulletin of the Peninsular Invertebrate Forum*.

Shortly after this I was contacted by a team from the Natural History Museum in London who also thought it would be interesting to look at the invertebrates in the Eden dome and, hearing that I had made a start, they invited me on board. The team, headed by Paul Eggleton, hoped to visit Eden once a year for the next five years. They wanted to look at the litter fauna and planned to do this by putting samples of the litter layer through a course sieve to remove large leaves, etc. and collect the fine material and animals in a mesh bag beneath. This is then placed in a Winkler bag that extracts the animals from the debris. These invertebrates were sorted into major groups (beetles, ants, bugs, spiders, etc.) and the collections were sent to relevant experts around the country. I received all the spiders. Most were money spiders that were common in woodland habitats, but one was a puzzle, a tiny, 2mm-long spider that was an orangey-red colour and had a pattern of tiny indentations on its abdomen. After a few enquiries it was placed in the tropical family the Anapidae. These specimens were dispatched to an arachnological colleague of mine, Rowley Snazell, an authority on this group.

Rowley identified it as a member of the genus *Pseudanapis* but was unable to match it with any known species, hesitantly suggesting it could be new to science. This caused great excitement at Eden, and we decided that if it were new, we would call it *Pseudanapis edeni*. Rowley set about describing the spiders, while I attempted to discover where these spiders had come from. I began by asking Eden where the plants from that area of the tropical dome had been sourced and chased up the suppliers. Unfortunately, the suppliers did not keep detailed records and, whilst they knew where their plants in general came from, the origin of individual shipments of plants was not recorded. The upshot was a list of possible locations around the world, which made it impossible to even start to guess the home of our spider. The paper was duly drafted and sent to the world authority Norman Platnick at the American Museum of Natural History. Norman gave it an initial thumbs up but said he would make enquiries and would get back to us. About one month later Rowley received an email from Norman saying, "I had a feeling I had seen this somewhere before and, after looking through my reference collection, realised this species has been recorded from Hawaii, where it was given the name *Pseudanapis aloha*." Norman thought it had been introduced to the island, as have so many other invertebrates. So, sadly, no new species from the Eden domes but still a curious global distribution – Hawaii and the Eden Project in Cornwall. I have a sneaking suspicion that it actually lives undetected in many of the large nurseries that supplied Eden, a feral spider that is being exported around the world with tropical trees. I am also sure that one day it will turn up in a sample of spiders from a forest somewhere in the tropics, finally revealing its true home.

Harvestmen are close cousins of the spiders, also having eight legs, yet all their body parts are fused into one giving them the appearance of peas with very long legs. They were always turning up in my samples, and it seemed only natural to add them to my list of arachnids to study. As my previous visit to the Field Studies Council Centre at Box Hill had been so successful, I decided to embark on another of their courses and signed up for a weekend with John Sankey, the UK expert on this group. Disappointingly only two students turned up, but the upside was that we had the full attention of our tutor. John was wonderful; he had us out in the field all day and taught us to identify many of the harvestmen with the aid of a hand lens.

John's enthusiasm was infectious. We had personal tutorials with him over cups of tea and long discussions on harvestmen biology over dinner. We visited his study where John showed us his invertebrate collections and examples of the species that we had not seen in the field. The study was an ex-army hut about 30ft long and lined with bookshelves. The air in here was filled with a distinctive aroma – the musty smell of many old books with a dash of alcohol from his spirit-preserved collection of harvestmen and spiders plus a generous helping of naphthalene from his boxes of pinned insects. This mixture of books and specimens is a heady mix and is instantly recognisable by anyone who has worked with entomologists. When I take students to the Natural History Museum in London, I always ask them to remember the distinct aroma on first entering the entomology department. I would often paraphrase Robert Duvall's famous phrase from the film *Apocalypse Now* – I love the smell of naphthalene in the morning.

I returned to Devon brimming with enthusiasm and added harvestmen to my list of invertebrates to record. With only twenty-four species in the UK, I had soon identified the common species that occurred in the South West. None was more abundant than the small, black-bodied, short-legged species *Nemastoma* that is characterised by two white spots on the upper surface of its body, a species so obvious that it was unmistakable. A fact that came to cause me great embarrassment. I can still recall going through some samples collected on the morning's field trip with a group of second-year students. One of them indicated a harvestman under her microscope and asked me what it was. Without looking down the microscope I pronounced it to be *Nemastoma bimaculatum*, a very common harvestman of woodland leaf litter, and moved on. The Fates were not to tolerate such arrogance. At the end of the lab session all specimens were collected into one container of 70% alcohol, labelled and stored for future use.

Two years later one of my students, Mike Hogg, had developed a keen interest in arachnids, so I set him looking at samples we had in store to gain some experience. Mid-morning he called me to his microscope with a puzzled look on his face. "This harvestman does not fit the key; it looks all wrong", he stated with a puzzled look on his face. "Can you tell me where I am going wrong? I could just see it under the microscope, believing that it was *Nemastoma*, so I sighed mentally and walked over to take a look.

The view down the microscope stopped me dead. The harvestman was black and had two white spots but also two golden spots at the rear of the body, which was covered in short, blunt-ended spines. We did not have any harvestman species that looked like this in the UK. I was startled, surprised, wildly excited that we clearly had something entirely new on our hands. Then the enormity of my error hit home. This specimen had sat unnoticed in my collection for two years. Recovering my poise, I approached the discovery with a new humility. I hastily plucked my copy of the book that dealt with European harvestman from the shelves, and we soon had a name for the genus; it looked very much like *Centetostoma,* a genus of harvestman previously recorded from the foothills of the Alps. Drawings were made and photographs taken before dispatching the specimen to Paul Hillyard at the Natural History Museum in London for an authoritative opinion.

The hedgehog harvestman *Centetostoma bacilliferum*.

We returned to the site and scoured it for further specimens, discovering a small colony of nine individuals which we left *in situ*. Paul quickly got back to us with the news that we had the correct genus but the species as such was unclear. It would have to be sent to the European expert in Germany for identification. It was, however, definitely a species new to the UK and possibly new to science.

Statements like 'new to science' are music to a biologist's ears because discovering undescribed species is the kind of thing that rarely happens in the UK. Mike and I were more than excited. We wrote a short note in the newsletter of the BAS to announce its discovery and to alert other arachnologists that there was a new species to keep an eye out for. We then waited for news from Germany. In the meantime, we searched the areas surrounding the original site of discovery and located two further sites for it within the city boundary. Paul produced one of his exceedingly fine drawings of the beast ready for the next version of his guide to UK harvestmen, but we still hadn't received any news from Germany. A year passed and Mike graduated, leaving Plymouth to seek pastures new, but still no news arrived about our unusual find. In fact, we waited for five years before we finally received news that the species had indeed been identified before and that it was originally discovered in the foothills of the Picos de Europa in northern Spain. So not an animal new to science, but the fact that its world distribution is restricted to only two sites in northern Spain and Plymouth, South Devon, remains inexplicable to this day[1].

Nothophantes horridus

While it is generally agreed that there are far more invertebrate species left to discover that we have so far identified, it is still an exciting moment when we find one that is new to science. If you work in the tropics, new species crop up all the time, but in temperate regions, such as the UK, it is a rare event. As you will have read in previous sections, hopes can be raised, but while a species may be new to the UK, a species new to science and discovered in Plymouth was so unlikely as to be a near impossibility.

Andy Stevens was the Nature Conservation Officer for the City of Plymouth and as such had been asked to conduct an invertebrate survey of an abandoned quarry that a developer wanted to turn into an industrial unit. As Andy had previously been a lecturer at Plymouth Polytechnic, he employed two recent graduates to conduct the field work and arranged for them to have access to the ecology labs at the Polytechnic. He also asked me if I would assist the two with the identification of the invertebrates that they would collect. So each day they scoured the quarry for inverts and returned to the lab in the late afternoon to identify them with some help and guidance from me. Andy would also take the occasional specimen home to confirm its identity. This worked well and a long list of common beetles, spiders, snails, woodlice and millipedes was generated. Then one afternoon Andy came into the lab with a tube containing a small spider. "I can't work out what this one is," he said. "It has a clear epigyne, but I can't find a match. Can you have a look?"

---

[1] Perhaps it is a relict species, with a fragmented post-Last Quaternary ice age distribution between the UK and Spain, like some plants and insects and mammals in southern Ireland that occur nowhere else in the British Isles, because the southern tip of Ireland was never covered by the ice sheets.

The horrid ground weaver *Nothophantes horridus*.

I looked at the spider under a microscope and thought I knew what this was but turning to the page in the spider book quickly realised that I was wrong. We then spent the entire afternoon checking every lead we could think of, but the identity of the spider eluded us. For further clarification, we sent the spider to Peter Merrett, the UK's leading expert on small money spiders. A week later he wrote back saying that he has never seen this species before and that he had sent it to the world authority at the museum in Paris. Wow! That made us sit down.

A month went by with no word, then Andy dropped by with a letter from the museum in Paris. They had not seen this spider before either; it was a member of the genus *Nothophantes* but could be a species new to science. What did Andy want to name it?

That was amazing! A species new to science from an old quarry in Plymouth! Andy talked to Peter Merrett about a name and, as the spider has a hairy abdomen, they resorted to their classical education and used the Greek *horridus*, meaning hairy. So it became *Nothophantes horridus*, the hairy *Nothophantes*. They wrote a paper describing the new spider from the female that we had and started searching for a male.

In the meantime, we were sure that this would stop the development of the quarry as this spider has been found nowhere else in the world. It could not be any rarer. The report was sent in, and we awaited the decision of the planning authorities. How naive we were. A few weeks later our hopes were crushed by a letter informing Andy that the development would go ahead. The quarry was bulldozed, the rock walls stabilised and a carpet of industrial units installed. We later returned to look for our spider on several occasions, but it has not been found on that site since.

As it had been found in an abandoned limestone quarry, Andy began searching other quarries in the area and finally discovered the spider in Radford Quarry which is on the eastern side of the river Plym. Here he discovered several females and the elusive male. He wrote another paper with Peter Merrett describing the male; now we could identify both sexes of this species. Tragically, Andy died shortly after making this discovery and we never did manage our planned meeting to show me exactly where he had found the spiders in Radford Quarry.

At the same time the insect conservation society Buglife had produced a list of common names for all the UK spiders, and *Nothophantes* became the horrid ground weaver. It remained in limbo for ten years until Andrew Whitehouse, the new Buglife officer for the South West approached me with the idea of applying for a grant to see if *Nothophantes* could be found in other quarries. The application went in and the money was awarded. We appointed Duncan Allen, an ex-Plymouth University student with a keen interest in spiders, to do the work. A plan was drawn up: Duncan would survey as many of the abandoned quarries as we could gain access to, including Radford, using a range of different sampling methods. As the spider had been found under stones, this would be his main focus, but he would also use a vacuum sampler to extract spiders from stony screes and drinking straws to insert into cracks in the cliff face as it had been suggested in the original paper that the spider could be living in such cracks in the rock. The idea was that if the spider was moving in and out of these fissures they might take up residence in the straws as these would offer a confined space of just the right size.

Duncan spent six months searching the quarries; on one occasion, we had an army of students in a search line sweeping Radford

Quarry and collecting every spider found under a stone. Yet, there was no sign of the horrid ground weaver. It was at this point that a developer who had obtained outline planning permission some years previously decided to apply to build fifty houses in Radford Quarry! The government was encouraging the construction of new homes and so the developer saw an opportunity. Why anyone would want to live on the floor of a quarry with 50m-high rock faces as the view was beyond me.

The application went to a public enquiry and Buglife fought valiantly to save the site. It was at the enquiry that the common name of the spider came back to bite us as the developer's lawyer kept referring to this horrid little spider. Despite the efforts of the developer, the enquiry ruled that permission to build the houses was denied. The developer then appealed and Buglife launched an online petition to gather support for the campaign and accumulated nine-hundred signatures. These were presented at the appeal; the application was once more denied. The site seems to be safe for now, but future pressures on the nation's housing stock may bring fresh challenges.

Around the same time John Walters and I had been asked to conduct a survey of an abandoned railway line adjacent to another quarry that was also earmarked for a vast housing estate. We visited the site six times between July and the following March and compiled a long list of invertebrates; they were all common and widespread.

On our last visit at the end of February, I turned over an old railway sleeper and spotted a small pink spider scuttling away. I grabbed it thinking it might be the pink six-eyed spider *Oonops*, a spider that is often associated with animal nests, but on my return home I discovered it was *Nothophantes,* I was stunned. This was the wrong habitat and the wrong time of year.

I called John and we both returned to the site and searched high and low but failed to find any others. We later returned on several occasions over the next few months to continue searching but to no avail. Then Sustrans, the UK walking and cycling charity, approached the Plymouth City Council who owned the old railway line with a view to transform the track into a cycle way that would help to link the proposed new town of Sherford to Plymouth City Centre. The conservation officer for the city arranged a meeting between Sustrans, himself and me to discuss the project and, although Sustrans were keen to proceed, they appreciated that it would be rash to do so until we had established where *Nothophantes* was located on site and how large the population was. They commissioned John and I to survey the initial section of the railway track for *Nothophantes* and, as we had discovered the lone individual in late February, we decided to concentrate our search over the late autumn and into the new year.

We split the work between us: I would run a series of pitfall traps that caught spiders walking over the surface of the ground plus a series of underground traps. The bed of the old railway line had been built of limestone chippings that were several feet deep; there would be lots of small spaces between the stones in which a wide range of invertebrates could live, including the horrid ground weaver. The subterranean traps would catch invertebrates that moved through the small spaces between the stone chippings. These traps consisted of a plastic bird feeder in which a small cup containing a preservative was placed. These feeders were buried in the chippings with the top at ground level. Invertebrates moving through the chippings would emerge into the space inside the feeder and fall into the pot. I knew these worked as a colleague of mine had used them successfully. He had been searching for a rare millipede that lived on shingle beaches. He spent the first week searching by hand and found only one, but the following week he buried two of the subterranean traps which were baited with slices of apple. Digging them out on the last day he recorded over thirty of the desired millipedes. This goes to show it's important to look in the right place using the right tool. I was optimistic that our spider was living down amongst the chippings.

John, for his part, would conduct a series of intensive hand searches of the area looking under everything that could be moved. I would empty my traps once a month, while John would conduct weekly searches. We sampled the site each month from November to February, but disappointingly my traps caught just a single horrid ground weaver. John on the other hand had found many individuals, discovering a large population that we now knew were active between November and February. This population appeared to be confined to a short section of the track that had been colonised by local trees, an area that offered a shaded and sheltered habitat. The problem was that Sustrans were keen to lay the cycle path through the centre of this area, which could have a serious impact on the horrid ground weaver population.

At this time, we had no information regarding the spider's biology or life history, so leaving the population undisturbed was the safest option. We discussed several possibilities, including raising the cycle path on stilts to reduce the effect that the cyclists would have, but the disturbance caused by the construction process was bound to cause serious damage to the area, so the raised path idea was dropped. Running alongside the wooded area was an old tarmac road that had been the main entrance to the old quarry when the railway was in operation. Diverting the cycle path onto this old road seemed like a perfect solution, but no one knew who owned this stretch of road. It seemed crazy not to explore this option, so the council began a search for the owner. Fortunately, they were tracked down and an agreement reached which allowed the road to be renovated and the cycle track now bypasses this important site. Plymouth is still the only place in the world where this spider is found, despite arachnologists across the UK hunting for it. John has continued to work on the project and is slowly unravelling the biology of this enigmatic species. Meanwhile, cyclists now hurtle along the new road totally unaware that they are passing just metres from one of the rarest spiders in the world.

Chapter 4

Diptera

Time's Fun When You're Having Flies[2]

For most people flies are bad news. We are all told at school how they paddle in cowpats to then stomp across your food just before vomiting all over it! Then there are the biting flies that can spoil our BBQs and picnics in the UK but in the tropics can spread a host of parasites and diseases. I can still recall a film I was shown in my first year at senior school detailing many of the tropical diseases transmitted by mosquitos and biting midges. I was both terrified and appalled. Flies were clearly very dangerous insects and hence the only good fly was a squashed one! But that's no easy task. If you try to swat a fly with a rolled-up newspaper you will know that most of the time this ends in failure. The fly peels away milliseconds before the paper sweeps in. The air being pushed in front of the paper generates a pressure wave that the fly can easily detect, and its multi-faceted

The hoverfly *Eristalis tenax* feeding on an Oxeye daisy.

---

2   Quote from the PhD thesis of Dr Gareth M. Prowse (2003). The insecticidal properties of a garlic oil, with special reference to its use against two dipteran pests. University of Plymouth, UK.

eyes have seen your movements in plenty of time to take to the air and avoid the incoming paper.

So, are flies the enemy or as the Monty Python crew could have said: "What have flies ever done for us?" As it turns out, quite a lot. They are the great recyclers, breaking down a vast array of organic materials from carrion through dung to plant materials. It's the fly larvae, the maggots that do most of this work, chewing the materials into smaller pieces and digesting them into less complex forms. These then become available to bacteria and fungi, which in turn make these nutrients available to plants. They are also important pollinators. It's not just bees that perform this invaluable task but a whole host of flies of all shapes and sizes. There are also many predatory flies that feed on other invertebrates and in so doing control their populations and, lastly, they are an important food for many vertebrates, many of which rely on flies and other insects as their food source. As a result, they are an important group of insects that, if properly understood and appreciated, can be seen as truly amazing creatures.

So exactly what are flies? We can all spot one the moment it invades our house, but what makes a fly a fly? The answer is: they have just one pair of wings. This group of insects has evolved from ancestors that originally had two pairs of wings, but the hind wings have been modified over tens of millions of years of evolution into small drumstick-like organs that sit just behind the current wings and act as biological gyroscopes. These modified wings are known as halters. It is these that enable flies to perform their impressive aeronautical feats, such as weaving and ducking out of the path of your flailing newspaper. They can also hover, fly backwards and even fly upside down.

Their halters are the main means by which they achieve this; these tiny gyroscopic appendages continually send messages to the fly's brain, letting it know if the fly has been shifted sideways by an air current or has begun to roll in flight. The fly can then adjust the movement of its wings to compensate and stabilise its flight. When it comes to being exceptional aeronauts, one group of flies stands out. I am sure many readers will have seen small black- and yellow-striped flies hovering over flowers on summer afternoons. These are hoverflies, members of the family Syrphidae. They hang in front of a flower while they feed on nectar, continuously manoeuvring against local air currents with the effect that they remain absolutely static, locked onto their food source. They hover to access nectar from flowers and hover to find a mate. In late summer I have often walked into a wood convinced that there is a swarm of bees resting in a nearby tree. The gentle hum rises and falls on the summer breeze, a gentle serenade that drifts across the afternoon. A hum that intensifies as I walk deeper into the wood, a deep resonant note pulsing through the trees. Hundreds of hoverflies hang evenly spaced among the trees, each locked onto a key feature, a twig, a leaf or just a shaft of sunlight. Each one barely visible as they float in the woodland gloom dancing in and out of the shafts of sunlight that fall from the canopy as they adjust their position. Each one almost silent, but as the numbers rise, so does the sound of their wings. Each is a male that has claimed a territory where it waits for a female to pass.

Hoverflies are wonderful ambassadors for this much maligned group of insects, an antidote to the negative reputation that houseflies and their relatives have acquired. Many hoverflies are brightly coloured, bearing wasp-like black and yellow stripes. Some are fury and resemble bees, while others are a shiny metallic blue or just plain black. This resemblance to bees and wasps is not coincidental; many hoverflies have evolved to mimic these noxious insects and, by doing so, gain some protection from keen-eyed predators, such as birds or reptiles. If a young predator has an unpleasant experience with a bee or wasp, it will tend to avoid anything that resembles them in the future, so the hoverflies cash in on this behaviour. These colourful adaptations also make them a popular group to study, and many naturalists have adopted them as their insect group of choice because they are attractive, easy to find and there are not too many species in the UK, a mere 276.

In entomological circles, flies (Order Diptera = Latin for two wings) have a reputation for being difficult to identify. I shied away from them initially but soon realised that, as we encountered them every time we took students into the field, I needed to get to grips with them. Having attended Field Studies Council introductory courses for other groups of invertebrates, I knew there was no better way to do this than to enrol in another one. That year just such a course was offered at Flatford Mill in Suffolk, the very first Field Studies Council Centre to be opened just after the Second World War. That summer I rolled up at the field centre to spend a week looking at flies. Flatford Mill is of course also famous for John Constable's famous oil painting *The Hay Wain* (1821), which is set on the ford next to the mill on the River Stour between the counties of Suffolk and Essex. The ford is long gone, but the mill and the small cottage (known as Willy Lott's cottage) that frame the scene are still standing. I was delighted to hear that I was going to sleep in Willy Lott's cottage during that week of entomological studies, residing in the set of one of the World's most famous paintings. Not a bad start.

The course was run by Dr Henry Disney who was then the warden of another Field Studies Council Centre at Malham Tarn in North Yorkshire and a man who knew a thing or two about flies. Each day would begin with a talk on some aspect of dipteran biology, then out into the field for a morning of observation and collecting. In the afternoons we would sort through the morning's catch learning the characteristics of some of the common families and attempting to identify a few specimens to species level. These were long days of counting wing veins, hunting for spurs on the flies' legs and, of course, examining a range of genitalia. After this intensive week, the Diptera were no longer as scary a proposition but a challenge waiting to be seized.

Having gained an insight into the diversity of British Diptera, flies became an important element of the Biology Department's ecology field trip to the Yorkshire Dales. One of the sites visited was Malham Tarn, a magical location set high behind the famous cliff of Malham Cove. The tarn itself is a dark lake lying across an upland bog that runs into the limestone pavement behind the cove. Woodlands stand uncertainly on its shores shielding the field centre that sits at the far end of the tarn. Dark, brooding, mysterious, even sinister on all but those good days when the sun shines from cumuli-flecked, blue skies. This bright, ultramarine backdrop then transforms the landscape to one of idyl; the warmth of the sun and the rippling reflections inspire local painters to come to the water's edge with palate and canvas. It is also a mecca for naturalists, its rich fauna and flora a powerful magnet. Each year we would bring our students here to study the lakeside habitats, but the big project was a study of the relationship between vegetation and the diversity of insects, and we used flies to illustrate this. Ecologists are interested in the way that the communities of plants and their forms affect the diversity of the fauna that live among them, so this was an excellent way to introduce students to the processes involved. We would select a series of different habitats along the tarn edge and then lay a transect of water traps. These were yellow dishes the size of a dog bowl, half filled with water to which a drop of detergent had been added. An unlikely way to catch flies one might think, but did you ever put out a yellow plastic paddling pool full of water overnight in the garden? The next day you would discover a carpet of drowned insects on the surface.

Nectar-feeding insects locate flowers using reflected light, some in the visible spectrum but also one in the ultraviolet range. Many man-made objects simulate these visual cues and attract insects, often with disastrous consequences for the insect. Dragonflies have been seen trying to lay eggs on shiny car bonnets, convinced that they were pools of water. Yellow paddling pools are mistaken for large flowers and attract nectar-feeding insects that realise too late the terrible mistake they have made. Their small mass means they are trapped in the meniscus that forms at the surface and drown. This confused behaviour has been utilised by scientists who study the abundance and diversity of nectar-feeding flies, hence the yellow water trap. These traps have proved to be an excellent way to sample not only nectar feeders but also their predators and parasites.

A line of traps laid out through woodlands on the edge of the tarn on a sunny morning in June would collect a large number of flies which were then collected into containers at the end of the day and preserved. These samples would keep our students busy for weeks on their return to Plymouth, spending many hours in the lab identifying the flies to family level.

This was an excellent way to introduce them to the huge diversity of flies that can be found in the UK. Over those lab sessions they would become familiar with many of the 103 families. These families can be tricky to distinguish from each other as at first glance many look very similar. The main characters that are used to separate them are the number and arrangement of the veins in the wings. Longitudinal veins radiate from the base of the wing to the outer edge with cross veins running between them forming enclosed areas known as cells. The position of these cells and their length relative to each other are also important discriminating factors when keying out the insects using published taxonomic keys. Students would soon learn how to tell their first and second basal cells from their costal or anal cells.

One of the most striking groups of flies encountered along the edge of the tarn were members of the family Dolichopodidae, the long-legged flies, which are predators of other flies. They fly low over the ground with a bobbing motion; their bright metallic-green bodies are stunning, but when viewed through a hand lens or microscope they have an alien beauty all of their own. On many of our days at Malham I have watched clouds of these flies crisscross the air at the edge at the tarn, a shimmering field of metallic points weaving an intricate network of sun-flashed tracers that were reflected in the surface of the tarn's shallows. An ephemeral symmetry of flickering gold and green in a delicate but deadly ballet performed by these copper-clad warriors in relentless pursuit of a mate or their prey. Like many insects, the males have extremely large genitalia at the end of their abdomens, indicating that in the world of flies, size does matter. These delectable dollies, as they are sometimes affectionately known, demonstrate that even flies can be truly beautiful and, like Luis Buñuel's *Bourgeoisie*, they display a discreet charm.

The Phoridae are another common group. These are tiny parasitic flies that have a robust and bristly appearance, giving them an aggressive look. I had been introduced to these on the fly course at Flatford Mill, where Henry Disney had referred to them as the horrid phorids. A little alliteration to embed them in our memories, but a name that is now widely used by dipterists. The phorids are surprisingly abundant in most habitats and are one of the first groups that students learn to identify as they possess a characteristically simple pattern of wing veins. Many attack a wide range of invertebrates, laying eggs on their host which then hatch and tunnel into their flesh, eventually killing them. But some only consume dead or dying invertebrates. One tropical species of phorid attacks ants, invading their bodies; when the fly larvae has fully fed and is ready to pupate it moves into the ant's head to do so. This kills the ant, but the final insult is that the larva then decapitates the ant and uses the empty head capsule as a protective case for its pupa.

I have also reared phorids from the egg cases of cave spiders where they eat the spider's eggs but appear to leave any spiderlings that have hatched unscathed. One species of phorid is well known for its association with human corpses, commonly known as the coffin fly. Gravid females will locate a freshly dug grave and tunnel down to the coffin where they lay their eggs. These hatch and feed on the corpse, pupating in the coffin. The emerging adult fly then makes its way back to the surface to mate and start the cycle again.

To most people flies are just flies. A few are well known, such as the crane fly or daddy long legs that blunders around our gardens in the autumn; houseflies we all know and horse flies can terrorise a walk or a picnic in the countryside, while with mosquitoes we are usually only aware they have paid us a visit long after they have departed with a sample of our blood. But some have an almost legendary status; one of these is the Scottish midge, tiny biting flies in the family Ceratopogonidae. The very mention of this midge strikes fear into the hearts of southerners, and the unwelcome arrival of these midges at Scottish beauty spots can send tourists running for cover with arms flailing.

I once spent a week on the island of Rum in the Inner Hebrides, a truly beautiful place. I was camping with an old friend of mine, Andy Kirkby. We pitched our tent by the loch side just metres from the sea on a patch of salt marsh that had the luxury of a tap. The hotel at Kinloch was just visible across the water, pine forests came almost to the water's edge framing the dark, cold waters of the loch itself. It was August and one of the sunniest spells that locals could recall in years; it should have been idyllic, and indeed it was until the wind dropped and the air became almost still, reduced to a mere breath that failed to stir the vegetation. At this point, drifting out of the saltmarsh, arose the dreaded Scottish midge, a fine, dark mist that knew we were there, knew we were fresh, available and, oh, so tasty. Homing in on a combination of our heat signatures and carbon-dioxide-loaded breath, they would overwhelm us in moments.

We spent the week cooking on an open fire. Fresh mackerel was in plentiful supply; we ate well but always with a cautious eye on the wind speed. By day two we had cut bundles of rushes from the edge of the salt marsh and stacked them by the fire, ready to throw into the flames. As soon as the wind died, we would toss the rushes on the fire generating a tall pillar of dense, acrid smoke; the midges duly came but stopped short of the smoke, so we would eat our meal dashing in and out of its refuge. We yo-yoed between not breathing and being eaten alive! Trying to eat while holding your breath is tricky, so we would take a bite of mackerel and leap into the smoke masticating furiously, attempting to swallow before we had to take our next breath, then leaping out and gulping a lungful of air. By the time we had taken a second breath the midges were on us, and the tail end of our inhalation had drawn midges up our nose. Spluttering, we leapt back into the smoke, only to repeat the cycle again and again until the mackerel was gone. Colleagues of mine working in the Outer Hebrides told me that they had resorted to wading into the sea each evening to eat their meals. Standing up to their shoulders in the water, plates held aloft in an effort to escape the marauding hordes.

During our stay on Rum, we attended a Ceilidh in the village hall, a fabulous evening of music and dancing with an occasional can of beer. Exhausted, we left the hall at around one in the morning and sauntered back to camp around the loch edge. It was not dark as the sun barely dipped below the horizon at that time of year. I remember it seemed so wrong as a southern lad to be able to see without the aid of a torch. The landscape had a slight golden tinge that drifted from the far horizon, high contrast and dark shadows made the path home dramatic. The light touched the edge of the forest, a mere dusting of photons but enough to reveal the interior as a dense black, impenetrable wall.

We were half-way home when the wind dropped. Even in the faint light available we could see the black mist rise from the loch side vegetation. We ran like the wind, dance fatigue forgotten, until our breath faded then slowed to inhale. Instantly a cloud of midges enveloped us. We accelerated and left them behind. As I escaped my personal cloud of followers, I noticed that Andy had a tear-drop-shaped cloud of midges hanging just behind his head; it was classic *Tom and Jerry*. I had seen this in cartoons since my childhood, but here it was for real. As he slowed, they formed a sphere around his head; he would then accelerate away leaving a pointed trail behind him. I looked over my shoulder and there, inches from my head, was my very own cartoon cloud. We ran, hearts pounding, adrenalin-fuelled, and made it back to camp in record time. Diving into the tent, we fought a rearguard action with insecticide spray and then burning insecticidal coils which generated clouds of toxic smoke. Choking on the fumes, we sat besieged by the dark mist wondering which was worse: suffocation by toxic smoke or being eaten alive by the midges. In the morning we swept the piles of bodies out of the tent, uncertain as to whether we were victors or survivors.

It's not just biting flies that cause us problems; some just love to move in with us and share our homes over the winter months. These are the infamous cluster flies. They aggregate in loft spaces, the gaps between window frames or any small space that is accessible from outside. If they stayed put it would not be so bad, but any disturbance sends them into a frenzy of activity. When I first moved to the edge of Dartmoor, they adopted the window in my son's bedroom. This always went unnoticed until that warm winter's day when Dave opened his window, resulting in a cloud of furiously buzzing flies erupting from the frame into his room. Dave fought back valiantly, rolled up comic in hand, but unfortunately this produced unwanted additions to his bedroom decor. The only real solution was to connect the flexible hose to the vacuum cleaner and systematically quarter the room. An hour later the room was clear and the walls washed, ready for next year's invasion. These flies overwinter as adults and look for a sheltered and dry location to escape from the winter cold. These could be holes in banks, hollow trees, under lose bark or any natural cavity. Our houses unfortunately mimic these conditions exactly. As a result, the cluster flies fail to differentiate between natural and man-made winter refuges.

It is not just one group of flies that aggregate in our homes; there are three very different groups that can be found taking advantage of our heating systems. The one that Dave had his annual battle with is the autumn cluster fly, *Musca autumnalis*. This is a similar size to, and a relative of, the housefly that we all know and love! The next has a bright metallic-green body and goes by the exotic name of *Pollenia rudis*, which is also a cousin of the housefly and is thought that the larvae are parasitic in earthworms, but little else is known of its biology. The third is much smaller, just two or three millimetre in length with yellow stripes on its thorax. This is *Thaumatomyia notata*, the yellow swarming fly. I have been called to examine many infestations by local environmental health officers or pest controllers; numbers of flies present vary wildly. I was lucky with my house as we would have a mere one or two hundred to deal with each winter, but I visited one house where pest controllers had removed ten black bin liners of the tiny *Thaumatomyia* from the loft! Once the cluster flies have discovered your house, they come back year after year; each new generation seems to know exactly where to find their winter accommodation. It is thought that each fly empties its gut while in residence over the winter; the droppings contain a chemical that acts as an aromatic beacon for the next generation which they home in on it. Insects have an incredible sense of smell, some being able to detect a single molecule of a chemical associated with the plants they feed on, members of the opposite sex or, as in this case, a safe refuge to overwinter in. It is also strange that these flies will select particular houses to move into. I have visited terraces of houses that look identical but one of them is covered in flies while the adjacent houses have none at all. Why this is remains a mystery.

I would often receive samples of flies sent in from local environmental health officers and be asked to comment on the likely source. Flies can be an enormous nuisance, but sometimes the perception is greater than the reality. If people have had a bad experience, they can become sensitised and react adversely to situations that would not normally bother them. This was the situation in the village of Tremar in Cornwall where a local poultry farm had mismanaged their chicken manure piles and generated a vast number of house flies which had invaded local homes. The problem was dealt with, but there were still regular complaints from residents. After two summers of complaints, the local environmental health officer asked me to determine whether the flies in the houses were from the poultry unit or the surrounding countryside. Hesitantly I agreed, expecting a legal challenge to my every move. We decided to use water traps to sample the flies inside houses (the same ones that I used at Malam Tarn). We set up these traps in a selected number of houses, at the local water works, and inside the poultry houses. Over that summer the environmental health officer collected the contents of the traps and dropped them into my lab where I identified and counted almost 2,600 flies. The data was conclusive: the flies in the houses were a random selection of fly species that were common in the agricultural landscape, the flies in the poultry house were those associated with chicken manure and the waterworks had flies with aquatic larvae that lived in the filter beds. A report was produced, and I breathed a sigh of relief.

In the early 80s a new head of the biology department, Colin Hawkes, arrived at the University of Plymouth and brought with him his research program on the agricultural pest the cabbage root fly. A suite of constant temperature rooms was quickly built to house the insect cultures that he would be working with and I was assigned to manage them. Several insect species were to be maintained, including housefly, white fly and the parasitic wasp that attack them and, of course, the cabbage root fly. These were exciting times with new technologies and a growing interest in entomology within the department. While being involved in the research was intellectually exciting, the maintenance of the insect cultures was less so. Cabbage root fly attacks plants in the cabbage family, the Brassicaceae. Eggs are laid around the base of a plant and the larvae hatch and tunnel into the roots on which they feed. One or two are not fatal, but if many larvae attack the plant, it loses its roots and dies. Therefore, to maintain a culture of root flies meant having a large root available for the larvae; swedes were deemed ideal for this.

Individual swedes were placed in pots of sand to await a batch of eggs. Meanwhile in separate large cages the adult flies were kept and fed a gourmet diet of honey and marmite sprinkled with a delicate dusting of brewer's yeast, plus a dish of sugar solution and water, each soaked onto wads of cotton wool, were also available.

Several hundred flies of both sexes were present in each cage, so the females quickly became gravid (loaded with eggs). To stimulate egg laying, a dish of sand on which sat a cube of swede was provided.

Every other day I would change the food and collect the dishes of sand in the cages. The sand was poured into a beaker of water and any eggs present would float off. These were then filtered using a disc of filter paper and the resulting cache of eggs were placed on a prepared potted swede plant. So far so good. Thirty days later the larvae had eaten their fill and tunnelled out of the swede into the sand where they pupated. My job was then to remove the rotting remains of the swede and drop the sand into a bucket of water so that the pupae would float to the surface. These were collected in a sieve and placed in a new cage to form the next adult population.

Behind a lot of good scientific research are many hours of dull, tedious work. Research is often perceived as a glamorous or even romantic endeavour, with the heroic scientist working against the odds in a bid to unveil the truth. In reality it is more routine. The old adage of Thomas Alva Edison that "genius is one percent inspiration, ninety-nine percent perspiration" is, oh, so true. The cabbage root fly project was a prime example of this.

Colin's team was interested in the fate of pupae that had overwintered in the soil after the brassica crop had been harvested. Farmers would rotate their crops, so where swedes or cabbages had been grown one year a cereal would be planted the next in order to avoid the cabbage root fly that was in the soil just waiting for the next crop of brassicas. We had to find a way of catching the flies as they emerged from the ground the year after the brassica crop. We came up with an inverted 9-inch flowerpot with a transparent plastic tube glued over a hole in the base. Flies emerging from the soil into the darkness of the flowerpot would see the tube as a single point of light to which they would be attracted. In the entrance of the tube was a small cone with a central hole through which the fly would pass. This cone would prevent the fly falling back into the pot. We would place our pots out in a field and visit them twice a week to count and collect any flies that had emerged. Sounds okay so far but scale up the numbers to allow a reasonable chance of catching a reasonable number of flies and it becomes a challenge.

We had five fields and each one would have between two and four hundred such emergence traps. Putting the traps out took a couple of weeks and checking them was a slow job, walking the line of traps bent double to see the tubes. It was not worth straightening as the next trap was just a pace or two away. As the tubes were often covered in condensation, each one had to be wiped before being checked for flies. If a fly was present, it had to be tubed and labelled with the date, trap number and name of site it had come from. Each site had to be visited twice a week, rain or shine … on the southern edge of Dartmoor it rains a lot. If the sun shone, it was great – a day walking in the fields, a picnic lunch in a shady gateway. I have wonderful memories of these moments. However, it was more often a slate-grey day with driving drizzle, full waterproofs and a snatched lunch in the damp interior of our Land Rover while the steam from our thermos flasks misted the windscreen.

The project was a great success and over the years answered many questions regarding the biology of cabbage root fly. We discovered that some pupae remain dormant in the soil for two years, catching out farmers who would plant their brassicas at one end of the farm one year and the other the next. The team also developed an analogue of the chemical that is given off by damaged brassica plants and which is the chemical cue used by the flies to locate the damaged plants. Small vials of this synthesised aroma were put into boxes lined with sticky glue which were then placed around the edge of crop fields. Incoming females found the synthetic chemical even more attractive than the real thing and flocked into the boxes rather than the crop. A grave error on their behalf, but good news for the farmer.

Colin Hawkes left to become Dean of Science at the University of the West of England in Bristol, which meant that the cabbage root fly research came to a halt. Even so, our understanding of the biology of this fly had been broadened and the synthetic analog of the swede's defence chemical had proven effective.

Chapter 5

Beetles, grasshoppers, crickets and other things

Beetles are the tough guys of the insect world. Encased in their armoured exoskeletons, they appear unassailable, and indeed in some cases are. While all insects possess a rigid skin that acts as an external skeleton, the beetles have taken it to the extreme and developed it into an impressive body armour that is far tougher than that of other groups of insects. For example, the tropical dung beetles in the genus *Dynastes* can support 850 times their own weight. I have seen people accidentally stand on one, but the expected crunch never came; they just got up and walked away. They have developed their exoskeleton into a suite of armour, evolving their front wings into strong, rigid sheaths that protectively cover their rear flight wings. Even though the flight wings are very delicate, beetles can still crawl into tight spaces that would otherwise damage their wings, thus staying out of reach of many predators.

My earliest recollection of beetles is finding stag beetles emerging from an old tree stump in my grandmother's garden. These were terrifying to my brother and me ... big, black and with

The Atlas beetle *Chalcosoma atlas*, a member of the subfamily Dynastinae (Danum Valley, Borneo).

those horrific jaws. We prodded them with sticks, fascinated by these strange creatures but scared that they would bite. My father reassured us that they were harmless, but even so we were always wary of these mysterious animals that appeared each year from the same old tree stump.

One of my other early memories of beetles also concerns my father. He rode a motorcycle to work every day. For those of you with an interest in such things it was a BSA Bantam, in standard post-war utility green. I thought it was brilliant and would often walk out along the country lanes to meet my dad as he returned from work on a Saturday lunchtime. I would stand on the verge and wave frantically when he approached to bag a ride home sitting on the petrol tank. One Saturday he arrived home a little shaky and with a bruise on his chin. Riding through the country lanes at speed he had collided with a May bug, otherwise known as a cockchafer; the impact had stunned him and he almost lost control, managing to stop just before he ran off the road into a ditch. The May bug became a legend in our house after that, a giant beetle that almost caused my dad to crash was a fearsome insect indeed.

One of my first professional encounters with beetles was when my then head of department, Len A.F. Heath, came to me with a problem. He was researching a disease of honey bees known as chalkbrood. He had been sent infected bees from across the country, but to his surprise beetles had emerged from some of these dead bees. Could I cast any light on what was going on?

I took the beetles back to my lab, put them under the microscope and reached for the beetle identification book *A Practical Handbook of British Beetles*, Volume 1 (1st edition 1932) by Norman H. Joy. Known as Joy for short, but in truth there was very little joy in this book. Mr Joy had produced his authoritative volume with a complete collection of British beetles at his fingertips; unfortunately, I had not. For example, when he states that 'the elytra are longer' or the 'hind tarsus is shorter' or 'the eyes are larger' it begged the question: larger or longer than what, and by how much? Joy could refer to his collection and compare lengths and widths, while I had to make assumptions. But if one made enough wrong assumptions, one eventually ended up with an answer that was clearly wrong and one began again and made a different assumption. After many wrong assumptions one would end up with the right answer. Joy was frustrating to an apprentice coleopterist but was also their salvation as it was then the only comprehensive key to British beetles available, it was a love-hate relationship. These days Joy is relegated to the top shelf in the bookcase as there are now many new identification keys that are far more user-friendly.

Having wrestled with Joy, I arrived at an identification: *Stegobium paniceum*, the drugstore beetle or biscuit beetle, so called as it infests both stored biscuits and stored pharmaceuticals – in fact, it was reported to infest almost anything. One source claimed it ate anything but cast iron. Now there is a thought, one that took me straight back to my boyhood comic *The Hotspur* that featured a story entitled 'The Beetles of Doom' in which a mad scientist had bred a species of beetles that ate steel and was using them to blackmail the UK government. Pay up or my beetles will eat the Forth Bridge, and they did.

Back to *Stegobium*. Its natural habitat is animal and bird's nests where it feeds on the organic debris that accumulates in the nest, including hair, feathers, skin flakes, the bodies of external parasites like fleas and lice, or the bodies of other invertebrates that invade nests. The beetles fly readily, and as they hunt for new animal burrows to breed in, the dark entrance of a beehive would be a natural choice. Finding lots of dead bees in the base of the hive provides a wonderful food supply and beetle eggs are laid on them, the larvae hatching out consuming the dead bees and often tunnelling inside them for safety. We wrote a short paper describing our findings which was my first time in print.

Ground beetles (Family Carabidae) are probably the most commonly encountered beetles after ladybirds. They are found under logs and stones. Most readers who have turned over logs will have seen them dashing off into the vegetation – sleek, black and fast as lightning. They come in a wide range of sizes, from 2mm to 2cm, and their colours can range from jet-black to brassy-greens and steely-blues. Their long legs mean they can move rapidly; they look dangerous and indeed they are. At their scale they are the tigers of the undergrowth. The 19th- century French entomologist Jean-Henri Fabre called them the 'frenzied killers of the night'. Their large, scythe-like mandibles are used like garden shears, slicing their prey into small pieces for easy consumption and it's not just other insects that can be sliced up. On many occasions I have encountered the violet ground beetle, *Carabus violaceus*, one of the UK's larger carabids and one that is fairly common. I have often picked one up to show students their impressive jaws only to have them slice into my fingers and draw blood, to the great amusement of my audience. However, not all ground beetles are carnivores; a few are vegetarian and eat mainly seeds. One in particular is *Harpalus rufipes* (rufipes, as it has reddish legs).

This beetle is famous as a pest of strawberry fields. They steal the seeds from the surface of ripening fruits, thus damaging them. It is also an ideal insect to introduce students to the concept of food selection. The students would offer the beetles different kinds of sizes of seed and note which the beetles preferred and ate. I would spend many evenings collecting *Harpalus* for them. As they are nocturnal,

A Carabid ground beetle.

this would mean a trip out after dark. The Polytechnic's horticultural station at Bere Alston, West Devon, was an ideal site to search due to the range of crops grown there and the environmentally friendly management regime conducted by the manager.

My son David often came with me to hunt for these small tigers by torchlight. Walking quietly around the fields, we would turn over logs, stones and pots to locate our prey and place each one into a separate tube. Dave thought this was great fun, out after dark hunting for tiny killers. I failed to tell him that they were vegetarians. Triumphantly we would return to the car with a dozen or so beetles. Driving out of the gate one night we were suddenly aware of something large in the hedge. We stopped the car and to our surprise a badger burst out of the hedgerow onto the road. It turned to look at us as if to say, 'What are you staring at?' and then trotted up the lane with the occasional glance over its shoulder. Cautiously we followed it in first gear. David was glued to the windscreen as neither of us had ever been this close to a badger before. We travelled with the badger for half a mile before it halted again and looked around one last time before walking under a gate and into the night.

One of the truly amazing behaviours that beetles exhibit is that of bioluminescence. Beetles don't have a monopoly on this ability, but one family of beetles, the Lampyridae – the fireflies or glow-worms, have made it their own. In the British members of this family, the females have evolved a system of emitting light from the end of their abdomens. They use this as a means of attracting a male to mate with. The female is flightless, sitting and waiting for males that can fly to come to her. In the UK there is only one native species of glow-worm. The series of flashes that the female produces are simple regular pulses of light as they do not have to compete with other light-emitting species. In other parts of the world, where more species are bioluminescent, more complex signals have evolved to ensure that each species finds the correct mate. Our glow-worm may have a simple signal but can still be extremely impressive. I recall taking an early girlfriend of mine on a romantic evening walk around Windsor Great Park and being amazed as the bushes and lower branches of the trees were decked with flashing lights. It was a mesmerising sight, as if the woodland had been strewn with fairy lights, their blue-green ambiance softening the woodland shadows. Slow pulses of cool light that lit the woodland path. A truly magical evening.

I also saw them in far smaller numbers in Cornwall when I lived in the small village of Landrake. I often took my two young sons on nighttime walks across the local fields. There was one hedgerow that always had half a dozen glow-worms flashing away to the delight of both boys. We took one home and kept it in a jar that lit their bedroom for a couple of nights before returning it to its hedgerow in the wild.

One of the most spectacular displays of bioluminescence I have seen was in Malaysia where beetles of the genus *Pteroptyx* put on a nightly display, a display so impressive that it has become an international tourist attraction, drawing people from all over the world. The beetles congregate in the mangrove trees that line the river at the small village of Kuala Selangor. Here the males kick things off by flying into the riverside trees and then flashing a signal at the females in the area. Large numbers of males congregate in groups along the river, generating thousands of flashing pinpoints of light that sparkle among the branches of the trees. On a good night it can be dazzling, a galaxy of tiny stars burning bright along the river bank. But the weather has to be just right to persuade the maximum number of males to take to the trees. Females are looking for the brightest flasher to mate with, so for an individual male to stand out it can be tricky amid the multitude of flashing beetles. However, if males synchronise their flashes, the brightest becomes more obvious. Hence the males present tend to slowly synchronise their flashes over the first hour or so and hope that they are brighter than the surrounding males. Females make their choice and the whole business starts again next night. This spectacle used to be far more extensive, offering a corridor of light that stretched from its current location to the river mouth. In fact, the locals claim that in the past fishermen returning to land would use the flashing lights as a beacon to navigate by.

Sadly, the numbers of fireflies are declining; the reason is all too clear. The larvae of these fireflies are snail predators and therefore prefer damp woodlands which have large mollusc populations. The fields that back onto the strip of mangrove trees which borders the river were once extensive wet meadows with a good snail population, but in recent years these meadows have been cleared and cultivated, or developed to build desirable riverside homes, a process which has drastically reduced the snail population. Loss of these wet habitats means fewer snails and, thus, less food for the larvae and therefore fewer fireflies. There is an ongoing programme to conserve the fireflies, but at the moment their fate hangs in the balance along with that of the local human community whose livelihood now depends on this tourist attraction. It would be a tragedy if this once famous colony of beetles were to vanish just because a few wealthy people wanted a riverside view. And an even greater tragedy if that choice of view rang the death knell for the local community.

How do the beetles make this amazing light? Generating light requires lots of energy and these insects are far from large. Light is normally a byproduct of chemical reactions that release energy in the form of both heat and light. These beetles have developed a system that releases most of the energy as light and a very small percentage as heat.

Some of the most bizarre beetles I have encountered were on the island of Borneo, a place that teams with strange and exotic insects. The most curious ones were the so-called trilobite beetles, members of the genus *Platerodrilus* from the family Lycidae. These are beyond strange, just downright weird. They are found on fallen trees where

A female trilobite beetle, genus *Platerodrilus* from the family Lycidae. Danum valley, Borneo.

they are thought to feed on the decaying timber and the fungi that is responsible for breaking it down, but this may not be the whole story as recent observations indicate that some may be predatory and could live in the forest leaf litter. They look like elongated trilobites, hence the common name, but can reach lengths of over ten centimetres[3]. The insects known as trilobite beetles are the females; the males look like ordinary beetles and are rarely seen. But here is the even weirder part of the story. The male hatches from its eggs as a larva, feeds and moults its skin until it is ready to pupate. After a short time as a pupa, the adult beetle emerges bright, shiny and with a brand-new set of reproductive organs ready to reproduce. As the adult male has wings, it can fly around in search of females. The female takes a developmental shortcut and, rather than entering a pupal stage, skips it and keeps on growing. It manages to produce a set of reproductive organs while retaining her larval body, so she is still a juvenile but reproductively active ... a trick known in biological circles as neoteny. The female ends up many times larger than any male, but as she has no wings she can only travel

---

3   By the way, if you are struggling to visualise this beetle, think back to the second ever Star Trek film *The Wrath of Khan* (apologies to non-Trekkies). In this film, Kahn (an evil intergalactic villain) places a mind-controlling creature in Ensign Pavel Chekov's ear, a creature that looks extremely terrifying and convincingly real, which is because it's a trilobite beetle; but just for the record, they do not possess any mind-controlling abilities ... as far as is presently known!

The veranda at Dunham Valley, Borneo.

as far as she can walk. So, while males can travel large distances to spread their genes, the females have to walk to the next fallen tree in order to ensure that the next generation has a good food supply.

The veranda at the Danum Valley Field Centre, Sabah, north-east Borneo, as mentioned before, is legendary; it adjoins the dining room and looks out over the surrounding rain forest. Tea and coffee are always available, and visitors often bring a bottle of something stronger to make evenings more sociable. If you sit and watch quietly, amazing animals will drop by. In the early hours of the morning, I have seen gibbons swing along the distant skyline calling into the dawn sky, freshly hatch mantids that look just like ants swarmed along the handrail and hornbills would often make their inelegant crash landings in nearby trees. Even orangutans occasionally passed by, halting their travels to rest in one of the large trees on the edge of the forest.

Returning to the veranda after a long day in the forest, we once found a violin beetle (Family Carabidae) sitting on the veranda handrail.

A violin beetle, *Mormolyce* species. Danum valley, Borneo.

These beetles have very flat and wide wing covers along with a very narrow body that protrudes forwards resembling a violin, hence the name. To be honest, there was an air of the ridiculous about it. How would this morphology be useful to a rainforest beetle? I was talking to one of the local research assistants later that day and he told me that their shape and colour mimic the seeds of one of the famous jungle climbing plants, the Lianas. This one, *Alsomitra macrocarpa*, produces gourds in the high canopy which split open to release many seeds that have thin papery wings.

These are excellent gliders which weave through the surrounding trees travelling significant distances on their way to the forest floor. Once there, the wings decompose leaving the seed that the violin beetle bears a remarkable resemblance to. A disguise that would offer protection from insectivorous birds and reptiles. They also have a second line of defence just in case a predator is particularly persistent. They possess a small gland at the end of their abdomen that can squirt butyric acid at any potential enemy and with deadly accuracy. If you have ever accidentally rubbed vinegar (acetic acid) in your eye, you will know how unpleasant this kind of organic

A seed of the liana *Alsomitra macrocarpa*.

acid can be. A beetle that looks like a large seed and can repel would-be attackers with a deadly spray ... now that is seriously impressive!

It was on the veranda that I experienced an encounter with Borneo's giant katydid (*Macrolyristes imperator*, one of the world's largest crickets). We walked out onto the veranda after another hearty lunch to find this giant insect sitting on the floor. It was huge, its size and bulk instantly commanded respect. We crowded around, admiring it and taking photographs. My old friend Steve Burchett picked it up for me to take a photograph and to show its size

Borneo's giant katydid *Macrolyristes imperator*. Danum Valley, Borneo.

compared to a human hand. The katydid overlapped his palm, and Steve has huge hands.

Not wishing to upset the katydid, we put it back on the floor and took a few more photos. By this time, it had had enough of the entomological paparazzi and began to stridulate. Now I can hear you saying "so what, crickets do that all the time. Surly, that's no problem?" But this was stridulation with attitude. The pitch was penetrating and the volume unbelievable. It took us through the pain barrier in milliseconds. Everyone clapped their hands across their ears and leapt several feet backwards. Stunned and reeling, we stared at each other in disbelief while the katydid turned around, gave us one last burst of excruciating sound and hopped off the veranda into the forest. Shock faded and laughter trickled through the group, we had just witnessed a devastating defence mechanism and, boy, was it effective. The giant katydid is not to be messed with. This had been another demonstration that Danum Valley is one of those special places where the beauty and complexity of the natural world is self-evident. I hope that it remains that way for many generations to come.

Before overseas field trips became fashionable, we would take the first-year students to Dorset for an introduction to field biology. We stayed at a hostel in Swanage and visited the local heaths, chalk downs and the dunes. Over the five days we examined the plant communities in all three habitats and the insects in two. On the heath the students investigated the ant communities of wet and dry heathland. While on the dunes, we conducted a mark-release-recapture (MRR) exercise on the resident grasshopper population. Ecologists are always interested in numbers. How many individuals of a species are present, or how many species are present, or both? MRR is one of the simplest ways to work out the size of a population.

It was a day-long procedure which began by marking out a 60-metre square with a measuring tape and canes, being careful not to step inside the square and disturb the resident grasshoppers. All

The common field grasshopper *Chorthippus brunneus*.

students would then line up along one edge of the square armed with a net to catch grasshoppers and a tube of quick-drying acrylic paint and a cocktail stick. They would move into the square en masse and catch as many grasshoppers as they could, marking each one with a small spot of paint on their pronotum (the first section on the thorax behind the head). Having made an initial sweep through, the students would then spend the next hour randomly walking in the square catching and marking as many of the remaining unmarked grasshoppers as possible. Often to grumbles of "How many more grasshoppers can there be?" It also became clear that some students were exceedingly good at spotting grasshoppers while others struggled to see them only when they jumped. A long lunch would then be called, and the assembled company would move onto the beach to eat, swim and doze in the sun for a couple of hours, while the marked grasshoppers dispersed and mixed with the unmarked ones.

While this may sound idyllic, the students' leisurely lunch was often disturbed by the local nudists. The area of the dunes that had the best grasshopper populations was on the boundary of the dunes that were set aside for naturists. So as students dispersed along the beach to swim or sunbath in a typically raucous student fashion, they would often disturb a family of naturists that had claimed one of the many sand dune hollows. This would result in the male in the group suddenly rising up out of the vegetation ready to defend their territory, hands on hips with a slight swagger and often a less than subtle pelvic thrust. I often wondered if an accompanying jump and a request to do the time warp might follow.

This posturing would leave nothing to the imagination and our students slightly aghast. Now entomologists are very familiar with the examination of genitalia in order to determine the species of the organism they are studying, but this was a well-known species that required no second glance, a set of genitalia too far, even for biologists. A rapid retreat would ensue, but young, fragile minds may have already been deeply scarred.

Despite the trauma of the genital displays, we resumed the exercise and re-entered the square to catch more grasshoppers, counting the number of marked and unmarked individuals that were netted. I know what you are thinking; if all the grasshoppers were marked, how could there be any unmarked ones to be caught? Well it's amazing how easy it is to miss grasshoppers amongst the marram grass. The thick, spiky tussocks are an ideal hiding place and the mottled appearance of the grasshoppers provide an excellent camouflage. So a fair number are always missed, even when there are lots of willing pairs of eyes and hands to search.

If the number caught the first time is multiplied by the number caught the second time and this total is then divided by the number of marked individuals in the second count, this will give an estimate of how many grasshoppers are present in the particular area sampled. This is known amongst ecologists as the Lincoln Index. The more you catch the first time the more accurate the estimate is. Students were always amazed at how many unmarked grasshoppers we caught in the second session (despite their best efforts the first time around) and how large these total estimates were. The bottom line is that there are an awful lot of grasshoppers in sand dunes and most of them are hard to spot.

Chapter 6

Insects, a new hope

We are at the dawn of a new era; the world human population is steadily rising and is predicted to reach over nine billion by 2050. That's an extra 2 billion mouths to feed, with still the same land area to produce this extra food. One could clear much of the remaining natural environment to gain more land for food production; however, we now know that we cannot survive without these wild tracts of our Earth. More especially the fact that the forests are the lungs of our world, absorbing carbon dioxide and producing oxygen. We need them.

The rise of industrial agriculture has kept pace with the growing world population over the last seventy years but has now reached its peak. We need alternatives. Genetically engineered crops, algal proteins and synthesised meats will all play important roles in providing this extra food. Another alternative is the use of insect protein as an additional food source, one that already makes a major contribution to human diets.

Many insects possess a very high protein content

The author tucking into a mealworm and cricket taco.

that is as good as or sometimes better than beef or fish. These protein-rich insects can be dried and ground into a flour to be used to make a wide range of foods for humans, and it is also going to be an important component of food for farmed fish, poultry and pigs. The oil that is pressed from it is high in Omega-3 oils and the chitin that comprises the exoskeleton has health benefits as roughage; recent research even indicates signs of antimicrobial properties. Plus, it shows promise as a biodegradable replacement for many plastics. Farmed insects produce large quantities of excrement known as frass in entomological circles. This is showing great promise as a soil conditioner and fertilizer.

The role of land for food production is one of the big issues of our time. Farming of insects requires very little compared to cattle, pigs and poultry. To produce 1 kg of beef you need 200 square metres (m^2) of pasture, 1 kg of pork requires 50 and poultry 45, while crickets just need 15 m^2. The feed conversion rates of insects are also more efficient: 10 kg of cattle feed generates just 1 kg of beef, whereas 1.7 kg of insect feed generates 1 kg of insect. In addition, insects mature faster than traditional farm animals and produce far fewer greenhouse gases. All these efficiencies add up to making farming insects a very attractive proposition. They are also great recyclers and can turn waste streams, such as restaurant and supermarket waste, into insect protein, a process known as bioconversion. This is destined to play an important role in emerging cyclical economies.

Before you start shouting that this is all very well, but you won't catch me eating foods made from insect proteins, it's too late. You have been doing it since you were born! Insects are abundant in all the crops we grow, whether they are pests or beneficial insects or just tourists passing through the crop. They are all harvested along with the crop and either contaminate or enhance the harvest (depending on your perspective). They are tenacious opportunists that also invade many of the crops we store, therefore any grains, beans or nuts always contain some insect remains. Even though there are strict laws to control the percentage of permitted insect particles, they are always there. Flour, coffee, chocolate and peanuts will all be enhanced with small amounts of insect protein. If you like figs, you will have consumed hundreds of tiny fig wasps that pollinate the flowers occurring on the inside of the fruit; but that's another story.

The consumption of insects, such as termites, grasshoppers and various caterpillars, by indigenous people in the tropics and sub-tropics is well documented. The famous Witchetty grub (larvae of the cossid moth, *Endoxyla leucomochla*), highly prized by Australian Aborigines, became a great tourist attraction on outback tours back in the 1980s and then achieved international fame as an object of gastronomic horror in the infamous television show *I'm a celebrity, get me out of here*, however the European perspective on this was that it was a last resort, falling back on insects as food when there was nothing else. Nothing could be further from the truth. Once local gastronomies were explored, it became clear that the often-seasonal harvest of insects was eagerly anticipated and consumed with great relish. In the late 90s, I remember reading an article in the Royal Entomological Society's magazine *Antenna* describing how the Bushmen of the Kalahari used a wide range of insects for food and as medicines. There are nursery rhymes in Malawi that celebrate the arrival of the latest insect harvest; rhymes that were chanted by children to encourage mothers to hurry up with the preparation and cooking of their favourite insect dish.

In 1992, the American entomologist Gene DeFoliart published a review of the range of insects consumed around the world and examined western attitudes to the practice. He concluded that Europe and North America viewed insect consumption as primitive, so when emerging economies began to westernise, they too would adopt this view and abandon insects as a source of valuable protein. DeFoliart wanted to raise the profile of insect consumption in the West in an effort to lose its primitive tag, thus encouraging those in the tropics to retain their insect-eating traditions.

These articles piqued my interest, but it was not until I was working on an exhibition about invertebrates at Plymouth City Museum that it came into focus. The week-long exhibition known as *Bugs* was part of a longer exhibition; Weird Worlds showcased the work of the science faculty at Plymouth University.

Our section of the exhibition looked at the role of invertebrates in nature, asking the question: "What would happen if we did not have woodlice, dung beetles, spiders and ants?" Each panel depicted the aftermath of each absence, and the final panel contained a leafcutter ant colony (see chapter 7. How one curious thing leads to another even curiouser thing). I was aware that we needed something different to pull people into our week on *Bugs* and was searching for a novelty. Then I recalled reading that the Liverpool Museum had organised a day where they offered visitors a chance to try eating insects; to the surprise of the organisers, it had been an enormous success. I phoned them and, yes, people had queued up to try the insects. In fact, the open day was so popular that they had run out. I asked where they had obtained the insects from, and they gave me the details of a company in London called Edible. I had my hook with which to draw the public in! This was brilliant, a real shock story to grab the public's attention.

There were to be a number of talks to accompany the exhibition and *Insects as food* had to be one of them, followed by the opportunity to actually try some. This would be the lunchtime talk on the Saturday that we opened our exhibition to maximise public and media interest. I started researching the topic and was amazed at how many insects are eaten in various parts of the world. Across southern Africa caterpillars of a species of emperor moth (*Gonimbrasia belina*,) that feed on the Mopane tree are collected, dried and sold as a tasty protein source. Wholesalers can sell them at $250 per kilogram. In fact, a wide range of caterpillars are collected right across Africa which supports a large caterpillar industry.

Pupae of the silk moth (*Bombyx mori*) are a by-product of the silk trade; roughly 300,000 tons are produced globally each year. Caddis fly larvae are a great delicacy in Japan where they are known as zaza-mushi in top restaurants. The fishermen who collect them own the collecting rights to sections of the river; there have been fights and brawls over boundary disputes as the catch is so valuable. Scorpions are popular in China. Water boatmen are prized in Mexico and grasshoppers of every kind are regularly eaten by people from Mexico to South-East Asia.

Termites are also highly prized in Africa and are often served in bars as an accompaniment to a cold beer. In the UK we have flying-ant days where all the ant nests in the area produce winged males and new queen ants, which leave the nest and fill the skies. The queens fly as high as they can and then wait for the fittest males to reach them and mate. Unsuccessful males fall to earth and litter our pavements and patios. Termites do the same thing, but winged male termites are much bigger and juicier than the workers. As dusk falls on the day of the termite swarm, the surviving winged termites are attracted to light, especially the brighter streetlights which are irresistible. Children place buckets of water under streetlamps and knock the flying termites into the buckets to catch them. They are so transfixed by the acquisition of these tasty treats that they are oblivious to the dangers of traffic and many accidents occur. Hence, the local councils now turn off streetlights in the termite season to keep children safe.

Why is it that insects are so widely eaten across the tropics while we shrink in disgust at the thought? If you wander into the bush in the tropics, it's relatively easy, once you know what you are doing, to gather enough insects to feed a family. In temperate regions you can spend all day and still not find suitable insects to feed even yourself. The reason is that we have seasons. Most insects in Europe grow frantically over the spring and summer but reproduce before the winter arrives, so they have just six months at best to grow in, often in far from optimum temperatures. This means that their growth is limited by the climate. In contrast, in the tropics there is no winter, just wet and dry seasons, so insects can grow for longer, with higher temperatures in which to accelerate their growth. Thus, some insects in these regions tend to be much larger.

Having researched the topic and assembled the talk, I phoned Edible in London and placed my order. All went well, but as time was short, I decided to take the train and collect them myself rather than rely on the post to arrive in time. On the Wednesday before the event I was walking around Farringdon in London looking for the Edible shop. The address had taken me to a series of back streets that were devoid of shops and none of the buildings appeared to have any numbers, so I started knocking on doors.

"Hi. Do you know where Edible is?"
"Who?"
"Hi."
"No!"
"Hi."
"Never heard of them."
Then eventually, "Yeah, that's us. Come on in."

I entered the room which was packed with boxes; there was a desk and a staircase. No shop. This was the Edible stock room. The guy who let me in explained that Tod (the owner) was not there but was sorting out their pop-up shop in Harvey Nichols in Knightsbridge, but he had left my order for me to collect. The package was surprisingly small; on reflection I guess it would be. Arriving back in Plymouth, I unpacked my box of delights. We had water beetles, BBQ-flavoured mealworms, chilled crickets, toasted ants, ant pupae, giant water bugs and ants in Belgian chocolate. A banquet fit for a king.

Saturday arrived. We set up tables, laying out the feast at the front of the lecture room. Each dish was labelled with what they were. We had cream crackers to spread the insects on and plenty of water to wash them down. We also had a large sign warning of allergic reactions and asking anyone with an allergy not to try them. A medical emergency was the last thing we needed on this first day of our exhibition.

We waited for our audience in what now seemed a vast empty space full of empty chairs. I was hoping for thirty people but dreaded that no one would turn up; after all, who would be crazy enough to risk being asked to eat insects?

A few people arrived. We exchanged polite conversations that echoed around the room. A few more appeared, then a few more, and suddenly people were pouring in; moments later the room was full. Museum staff were beginning to turn people away; I couldn't believe it.

The talk went well. Lots of 'Ohs' and 'Ahs' and a sharp intake of breath when I pointed out how insects contaminate our food. There was even applause at the end. We then launched into the main event. I made an announcement about allergies and explained what was on offer. In true British fashion, the audience formed an orderly queue. We asked them to try just a few of what we had so that everyone could have a taste. People filed slowly through, fascinated by the food on offer.

There was lots of pointing and nudging, tentative sampling and surprised smiles. I noticed that some folks were on their second helping. One small boy was on his fourth circuit. He had developed a liking for the giant water bugs and was working his way through the entire tin. Over the next twenty minutes he ate them all! No one else dared to try them.

As the room emptied, many people stopped to talk to Karen, Helen and I to say

The boy who loved to eat the giant water bugs.

Insect snacks offered at the Weird Worlds exhibition.

how much they had enjoyed the opportunity to try insects and how surprised they were. The crowd dispersed and we were left looking at a table of empty plates. Who would have thought it! A great success and, of course, many of our diners had gone on to look around the exhibition.

Helen, the curator of Natural History, approached me. "We have had so many disappointed people that could not get in. Could we do it again next Saturday?" After a quick discussion with my co-organiser Karen, we agreed. I made a desperate dash to Edible's London store to obtain more supplies, and next Saturday was indeed another success. This was so unexpected. I still can't believe it. I now have a reputation as the man who eats insects.

Insects as snacks were great, but I could not help wondering what they were like in a meal. The opportunity came my way sooner than expected. I was organising a one-day conference on Cave Biology. One of the speakers was Dr George McGavin, then Curator of Entomology at the Oxford Museum of Natural History. I had read an article in *The Guardian* newspaper in which he had cooked a dish of crickets and vegetables for their journalist. We were looking for a meal after the day's talks, so could this be what we were looking for?

I contacted George, and to my surprise he eagerly agreed to cook an insect-based meal after the conference. All I needed to do now was book one of the University's small dining rooms and obtain permission to use the associated kitchen. Simple, what could go wrong?

I approached the catering manageress. Apparently, this was not going to be quite so simple after all: the dining room was fine, no problem, but the kitchen! There was a sharp intake of breath. You are going to cook what?

After much negotiation, the manageress agreed to let me use the kitchen as long as we brought our own woks and utensils and ensured there was not a trace of our visit when we left. This was to be clandestine cooking at its most invisible. We had to be culinary shadows as the chef must never know we were ever there.

George agreed to work out a menu and gave me the name of the company he considered to be the best cricket farm in the UK: Monkfield Nutrition. A week before the meal I ordered a sack of crickets which arrived on my desk at the university a few days later. A jolly, chirping parcel that had the mailman scurrying rapidly out of the room and generated raised – but resigned – eyebrows from my colleagues who shared the office with me.

To my wife's horror, I took the insects home to process. There were black crickets, locusts and meal worms that had to be sorted, killed and sterilised. My wife shut all the doors between the kitchen and the lounge and forbade me to emerge until all the insects were in the freezer. The process was to pick out all the very active individuals and leave any moribund ones behind. The active ones were placed in the freezer to kill them; once frozen they were blanched in boiling water to sterilise them and then re-frozen until they could be cooked. The five hundred crickets arrived in a sack filled with egg boxes. I would need something to put them in to sort out the ones I would be using. A large deep plastic tub seemed ideal. I emptied the sack into it.

Instant chaos.

I had just set off an orthopteran bomb.

Crickets were jumping everywhere.

I slammed a lid on the tub, but a tsunami of crickets was already sweeping across the kitchen. I spent the next half hour chasing crickets and finally retrieved most of them. There was just the small matter of those that had retreated under the fridge.

Maybe my wife won't notice.

I started again with a larger tub, taking one egg carton at a time. This worked well and I dealt with all the crickets, then the meal worms, which were far less excitable, and finally the locusts. By then I thought I had everything under control. A last inspection of the freezer and I retired to the lounge via doors I had not seen shut since we moved in.

The next morning breakfast was the normal busy affair as we got ready for work. I left the kitchen to grab some papers I needed at work that day when there was a call from the breakfast table.

"Schat!" (Pronounced shaaht, a Flemish endearment meaning 'treasure'.)

"What is this?" my wife exclaimed. Then a few seconds later

"Schatje!" (Lesser treasure.) "Please remove whatever it is."

I rushed back to the kitchen to find a locust sitting on the edge of my wife's cereal bowl.

"It's probably just hungry," I said nonchalantly, but there was no redeeming myself from the situation. I placed the said insect in a container and took it to work, making a hasty exit.

Despite our breakfast guest, I was not in the divorce courts and the conference went well. We had a talk on aquatic insects in Yorkshire caves and another on parasitic flies that live on bats. I talked on what cave spiders have for lunch and George gave a humorous account of a joint four-week expedition with the Royal Marines to explore a cave system on the border of Thailand and Burma. The meal that George had planned was grilled locusts on a bed of mashed avocado, stir-fried veg with crickets and oyster sauce plus meal worm satay and noodles. This was followed by crickets that had been marinated in rum and then dipped in chocolate. These were scattered onto ice cream. All this had to be prepared and so a small team invaded the university kitchen to cook the meal.

A dozen covers were prepared for the day's speakers plus Robin Wootton as the local Royal Entomological Society representative, four students and my wife. Everyone enjoyed the meal, even my wife who got to re-acquaint herself with our breakfast guest, but this time she had the upper hand. A day of both culinary and entomological success plus, I think, a first for both the Royal Entomological Society and Plymouth University. We then undertook a forensic clean-up of the kitchen and, as far as I know, the chef had no idea that we had ever been there.

The Royal Geographical Society

South Kensington is one of London's cultural hotspots; the block that is bordered by Hyde Park and Cromwell Road represents a concentration of intellectual institutions that explore, catalogue and interpret the world around us. There is the Natural History Museum, the Science Museum, Imperial College, the Victorian and Albert Museum, the Royal Albert Hall and even Baden-Powell House, the HQ of the Scout Association. But as I walked the sodium-lit streets on the edge of Hyde Park on that late October evening, it was none of these that I was heading for. My destination was that legendary institution synonymous with exploration and adventure, the Royal Geographical Society (RGS). What, you may well ask, was a biologist doing at the RGS?

Life is full of surprises. When I received an email inviting me to give a talk at a meeting on the theme of 'Feeding the Nine Billion', my eyebrows did rise a little. I am not a geographer, but it was my entomological knowledge that the RGS were keen to tap into. How real was the prospect of using insect proteins to feed a growing world population? That's what they wished me to talk about. The meeting would comprise a panel of three experts and a chairman; there would be three short talks to introduce the topic followed by a discussion with questions from the audience. The chair would be Jay Rayner, the novelist and TV food critic, while the other panel members were Tim Wheeler (Deputy Scientific Advisor and Director of Research & Evidence Division at the Department of International Development plus being professor of Crop Science at the University of Reading) and Edd Colbert (Campaign Coordinator for the Pig Idea). Tim was there as a government advisor on food security and climate change, while Edd's organisation was campaigning for the recycling of food waste as pig feed (just as we did when I was a boy). Between us we had to sum up what we knew regarding the prospects of feeding the growing human world population and then engage in a discussion with the audience. This sounded like a wonderful opportunity to bring the idea of insects as food to the attention of a wide cross-section of the population, so I accepted the RGS's invitation.

Hence, I found myself walking along the bustling pavements on the edge of Hyde Park on a warmer than expected October evening, heading for the RGS HQ. I arrived at the reception and perused the exhibition of stunning photographs of Eastern Europe while I awaited my host. These images were amazing. Shot from the air, they presented that exotic view that we so rarely have the opportunity to see: images blazed with crisp symmetries, abstract patterns imbedded in the landscape and intense colours. I wished I had longer to explore these images, but my host suddenly appeared. Amy Lothian introduced herself and led me to a room where Edd was waiting; introductions were duly made. There was an air of tense excitement. Edd was telling me about the meeting he had just come from at the London Mayor's Office regarding an event that the Pig Idea were planning in Trafalgar Square. Jay arrived, a storm of enthusiasm, fresh from his part as the Devil in the first British performance of Frank Zappa's *200 Motels* the previous night. Tim arrived last, having rushed from a meeting with government ministers. I began to feel that just coming up from Plymouth was a little pedestrian!

At the Royal Geographic Society (from left): Jay Rayner,
Tim Wheeler, Peter Smithers, Edd Colbert.

We were miked up and waited for the audience to settle down. Amy gave us the nod and we walked past the statues of Shackleton and Livingstone, who gazed sternly down as we passed into the famous Ondaatje Lecture Theatre. A procession of four that hushed the conversation buzzing around the room. The theatre's oak-panelled walls exude a sense of history, adding an air of gravitas. The auditorium was not full, but there were over four hundred people present and the row of red leather chairs were waiting for us on the podium, picked out by bright, unforgiving lights.

We settled into our seats and the auditorium fell silent. The director of the RGS, Dr Rita Gardner, welcomed everyone and introduced the evening and Jay Rayner who then introduced us and the topic. Jay talked eloquently and passionately about food production in the UK and the security of our food supply. He linked unrest in the Middle East and the Arab spring with rising food prices and shortages, leaving us with the thought that we are just nine meals from crisis should the UK food supply chain be interrupted. Just how long it would be before riots erupted on UK streets once the supermarkets were empty. Tim Wheeler then discussed the role of new agricultural technologies. Could these solve the problem? He ended on the note that there was no prospect of this happening; there was no silver bullet. We urgently need to explore other avenues.

I then talked of how insect farming could generate large amounts of animal proteins far more efficiently than farming vertebrates as we do now, explaining how this could be extremely sustainable if we utilise bio-waste streams to feed the insects. I also explained how two billion people already consume insects as a regular part of their diet and eagerly anticipate these tasty treats and went on to discuss the recent UN report on insects in human diets. I concluded by saying that we need to shift our perception of insects from one of disgust to one of delight, following the path that the UK had already taken in embracing sushi.

Edd Colbert discussed the use of processed food waste as feed for pigs rather than sending it all to land fill sites. He outlined the problems that had arisen in the past and the processes that now exist to ensure that food waste does not pose a hazard in the animal feed chain.

For the next forty-five minutes we fielded questions from the floor dealing with a wide range of topics. These ranged from water security, alternative foods in the USA, insects as animal feed, and just why can't we feed food waste to our pigs at the moment? The evening was concluded by the director who thanked everyone present and announced the topic of the next event, which would be Big Data. Members of the audience then approached the podium with questions that had not been aired and many useful contacts were made.

While feeding nine billion people had been the focus of the evening, there was a sudden downsizing of our perspective and feeding the four became a pressing issue, so the RGS director and her staff escorted us to a local restaurant.

'Feeding the Nine Billion' had been a wonderful opportunity to take part in an event hosted by a legendary institution.

Hosting events that raised the profile of insects as food became a regular occurrence. I gave talks to the Plymouth branch of the Children's University on several occasions and, of course, one was on insects as food. The talk went well and afterwards I offered the audience a large plate of crickets on cocktail sticks. The crickets had been flash fried and then dipped in either soy or sweet chili sauce. The children crowded around the table enthusiastically picking up cocktail sticks and daring each other to try them while the parents hung back, egging their children on. My plate of skewered crickets was going fast, so I offered the adults a chance to tuck in before they all went. Curiously they all remembered that they had eaten just before coming to the talk and really should not have anything else.

I ran another such event as part of National Insect Week in Yorkshire. Here I worked with a local chef, Lionel Strub, to prepare some insect-based snacks for local school children. After I gave my talk, Lionel would cook up some crickets and meal worms. He was going to produce an insect stir fry, but just to get everyone going, he prepared a tray of blinis on which he spread goat cheese topped with freshly cooked cricket. I offered these around the room to a forest of eager hands. There was then a noticeable pause as hesitant children tentatively bit onto their first insect snack. This uncertainty was followed by surprised smiles, rapid crunching and a request for more – all except for one boy who was obviously greatly distressed. His face was screwed into a mask of pure horror as he spat the blini into a paper cup. I sat down next to him.

"Are you OK?"

"Yes," he said. "But that cheese is disgusting."

Working with chefs to produce insect-based meals was fun but made me wonder if there were any restaurants that served meals featuring insects. There were a couple of London establishments that had insect dishes on their menu occasionally but none on a permanent basis. That is until I heard about Grub Kitchen in St David's, Pembrokeshire. It was at the launch of National Insect Week in 2014 that I met Sarah Beynon. She was setting up an entomological attraction and research centre on what had been her grandmother's farm. This was to be Dr Beynon's Bug Farm, a museum, education and research centre, a working farm and a café run by her husband Andy who is a chef. She then casually slipped into the conversation that he was developing a menu which featured insect-based meals. I was intrigued and had to know more. I arranged to visit the embryonic project and talk to Sarah and Andy about their plans and aspirations. My wife came along to take in the dramatic scenery of the Pembrokeshire coast and share the driving.

It's a long drive from Plymouth to St David's, so we stayed overnight in Bristol. The next morning we set off across the Seven Bridge. The motorway hauled us through the urban landscape of South Wales and onto the more rural A40 to Haverfordwest, where there is a sudden change of scenery as the landscape opens out. Leaving the town on the St David's Road, we wound across more open country, the horizon began to lower and the view was wilder and more windswept. Descending into Newgale, the road was almost on the beach, the sea now a constant companion as we travelled on to St Davids, the UK's smallest city. We threaded our way through its narrow streets and took the road north, out into the open landscape once more. We were looking for a sign; we knew the farm was a short drive from the city but were not sure exactly where. Driving slowly, we scanned the roadside for clues; then we saw it, a small placard bearing the silhouette of a dung beetle, a sign that could easily go unnoticed, whereas to an entomologist it was unmistakable.

Turning into the drive, we edged along the unmade track to the farm, parking in the courtyard where we stood in the sunshine and awaited our hosts. Looking around, it was apparent that the farm had seen better days, which only highlighted the scale of these two entrepreneurs' ambitions. Andy emerged from one of the buildings with a warm welcome. Sarah had been delayed and would join us soon, but in the meantime, would we like coffee? He escorted us to the barn that had been recently renovated. Coffee was made and Andy explained that the barn we were in would be the classroom, the other buildings around the courtyard a museum, an insect zoo, research facilities and a shop. He then took us into a room that was bright with freshly painted, white walls. This would be the café's kitchen and the partially collapsed shed that adjoined it would be the restaurant. I could not help thinking that there would be years of work to be undertaken here; these two were raving optimists.

Sarah arrived bringing her wild enthusiasm to the conversation, in stark contrast to Andy's calm and measured manner. Andy had prepared a light lunch of his now famous Bug Burger with polenta chips. These were amazing, rich meaty and bursting with flavour. My wife could not believe there was no meat in them. The polenta chips were also divine. Having tasted the potential that Grub Kitchen had to offer, we then discussed the ideas and philosophies behind it. Grub would offer a wide range of sustainably sourced foods to local diners and visitors to the Bug Farm. All raw ingredients would be locally sourced or foraged and some would be grown on the farm. The plan was to offer traditional meals with a slight twist and, while many diners would shy away from an insect-based meal, Andy hoped to tempt them with a selection of insect-based sides and tasters.

Andy and Sarah have come together from opposite ends of the gastronomic spectrum. Sarah is passionate about food production

Andy Holcroft and Sarah Beynon at Grub Kitchen with
Andy's signature bug burger and polenta chips.

and aims to shift farming attitudes towards sustainability, while Andy is concerned with the way we consume these products and wants to educate his diners to be more discerning as to the provenance of their meals. The combination of these leads to an holistic approach to good food. I left them to their Herculean task, and over the next year Andy and Sarah updated me with regular progress reports. Then, in 2016, I was invited to the Grand Opening of the Bug Farm, a ceremony that coincided with National Insect Week (NIW).

My wife and I arrived to find the old farmyard buzzing with a wide variety of guests. The local Mayor and Sarah's parents were there. Prominent entomologists and the author M G Leonard were also among the guests. I was amazed at the changes to the place; the once rundown buildings I visited just two years ago were indeed now museums, labs and classrooms. The kitchen gleamed with stainless steel surfaces and the derelict shed that was full of old agricultural machinery was now a smart, inviting restaurant, its walls hung with local artwork. The museum was packed with dramatic images and

Cutting the ribbon at the opening of The Bug Farm.

fascinating information, while the zoo offered a range of colourful invertebrates presented in attractive settings.

The local primary school was also here on a bug hunt in the farm's meadow, as part of the NIW, followed by a reading by M.G. Leonard from her recent book *Beetle Boy*. A red ribbon that had been stretched across the entrance to the yard was cut by a small army of people who had been involved in the farm's genesis. It was a fabulous day.

Andy has meanwhile supplied insect-based buffets for several entomological meetings. The first was the Royal Entomological Society's annual meeting, held at Trinity College Dublin in autumn, 2015. He laid on a sumptuous feast at the President's wine reception. This banquet of delights featured mini versions of the famous Bug Burger, savoury crackers adorned with cricket hummus topped with a mealworm or cream cheese, celery topped with a scattering of ants plus Andy's chocolate chip cookies made with cricket flour. The Provost of Trinity College and the Society's President took the first hesitant bite, boldly going where few entomologists had gone before. Hesitant anticipation turned to smiles of surprised appreciation, a reaction that sent a clear signal to everyone else whose caution was now overcome by curiosity and who rushed to sample these novel delicacies. To Andy's delight the table was swept clean, just a handful of broken crackers remaining.

In 2019, Andy laid on an even more impressive buffet at the Society's annual meeting of the special interest group (SIG) on *Insects as Food and Feed*, held at the Royal Agriculture University at Cirencester. This lunchtime buffet produced on both full days of the meeting may have by now been 'preaching to the converted' but it was eagerly appreciated and again swept clean.

Grub Kitchen is rapidly gaining a national reputation with chefs such as Michel Roux Jr. dropping in to try Andy's novel approach to sustainable dishes and endorsing his approach. Never a couple to stand still, Andy and Sarah have now evolved a new enterprise out of Grub Kitchen: Bug Farm Foods offers a range of insect-based products for sale, which includes whole insects and insect-based cookies but also their new product Vexo. This is a meat substitute made with plant and insect proteins, a product that has been tested by the most critical members of our society … school children. Vexo has been offered as part of the regular menu in Welsh schools and has been well received. Many children in this trial selected Vexo dishes over the more traditional dishes, indicating that younger members of society are far more open to new foods than adults.

The rising interest in insects as food and feed led to the Royal Entomological Society setting up a SIG to bring interested parties together and promote the exchange of ideas. *The Insects as Food and Feed* SIG now holds an annual conference and is attended by entomologists, insect farmers, animal feed producers, restauranteurs, vets, lawyers and major retailers. The meeting offers an annual overview of this rapidly expanding industry.

Andy's Ento buffet at the Ento15 conference, Trinity College, Dublin.

Andy's Ento buffet at the Insects as Food and Feed conference, The Royal Agriculture University, Cirencester.

Insects are destined to play a vital role in both animal feeds and as a meat substitute but will also offer new gastronomic adventures to diners who are willing to engage with this new challenge and explore new culinary horizons.

Chapter 7

How one curious thing leads to another even more curious thing – bug hunts and exhibitions

When I first moved to the small village of Horrabridge on the western edge of Dartmoor, I was asked if I would visit the local cubs at their summer camp and talk about bugs. This was a little daunting, the idea of talking to a field full of small boys who were away from parental control and running amok filled me with terror. I had been a cub leader when my boys were that age and can still recall the high-octane anarchy that a cub camp can be. But steeling myself, I assembled a short talk that I thought might appeal to small boys … a talk that was full of slashing jaws, slurping digestive juices and erupting parasites.

I arrived at the camp a little nervous and was offered a cup of tea by the cub leader Arkela (for those of you who have not read Rudyard Kipling's *The Jungle Book*, Arkela is the leader of the wolf pack; if you did not know this book read it now. It's not at all like the Disney cartoon). Now this tea was good stuff; tea made with water boiled over a wood fire possesses a particular flavour, a flavour that took me straight back to when I was a boy scout. Just one whiff of this nostalgic aroma and I felt right at home. Arkela offered me a chair in the middle of the field, and the boys and attendant adults sat in a semi-circle around me. I launched into my tales of predators and parasites, and time slipped by. As I came to the end, I realised that my 20-minute talk had lasted almost an hour, but the circle of excited faces still sat there eager for more. It had gone extremely well with lots of good feedback; some of the parents were horrified, but the cubs and their leaders loved it. I had also really enjoyed the experience.

When the leader of our local Wildlife Watch Group, Jonathan, asked if I would lead a couple of bug-related events for him, I was only too pleased to do so. The first was a bug hunt around a local meadow where we hunted for insects and spiders with nets and umbrellas borrowed from Plymouth Polytechnic. This was also the birth of my bug numbers talk which turned out to be a success, and I have used it many times since. It was a guessing game where I would call out a group of animals and the children guessed how many different kinds there are on the planet. I would start with mammals and hands would shoot into the air:

"100!"

"300!"

"2,000!"

"No, its 5,487 species of mammals that share our planet with us. How about reptiles?"

"3,000!"

"6,000!"

"No, it's 8,734. What about fish?"

"10,000!"

"15,000!"

"31,000!"

"Close, it's 31,153. Plants, how many do you think there are on our planet?"

"50,000!"

"100,000!"

"200,000!"

"Nearly, it's actually 281,621. But what about the insects?"

"500,000!"

"1,000,000!"

"10 billion!!

"Infinity!!!"

They were by now number crazy. I would try to restore order. "1,000,000 is the right answer, well done!"

The the number of insects that scientists have discovered and given a name is now 1.2 million, but we know there are lots more to be found and, indeed, hundreds of new species are still being discovered every year. The American entomologist Terry Erwin was one of the first to try and work out how many we have left to find and came up with 30,000,000 species of insect yet to be discovered. This number sparked lots of debate and many recalculations, but no consensus was ever reached. The best estimate that is generally agreed upon is that there are 'probably' somewhere between 5 and 8 million insect species yet to be discovered. Terry Erwin then came back into the fray saying, "I hold up my hands. I got it wrong. It should be 60,000,000." In fact, the only thing we know for sure is that there are an awful lot of insects that we have never seen.

Following the great success of the bug hunt, we organised a morning looking at freshwater invertebrates in the River Walkham, near Yelverton in Devon. We chose the very picturesque Magpie Bridge where the river was only a foot deep and tumbled over a bed of small to medium-sized stones, ideal for freshwater insects and easy for small entomologists to access. Here we were looking for mayflies, stoneflies, caddis flies and water beetles. The standard method of looking for these insects is to stand in the river with a fine mesh net in a metal frame on the end of a wooden pole. This is known in the trade as an FBA net as it was designed by the Freshwater Biological Association. Standing upstream of the net, you kick the river bed to disturb the loose stones and any insects that are displaced are washed downstream into the net. The net is then emptied into a tray of water and the catch inspected. The Environment Agency (EA) has teams doing this all over the country as it is possible to assess the health of a river by noting which aquatic insects are abundant and which ones are not. Any pollution incident will deplete populations of the more sensitive insects. A quick count of which groups of insects are present and how many of them there are, therefore offers a rapid method of checking water quality.

The EA teams standardise their collecting method by sampling one area of a river for three minutes. Standing on one leg in a fast-flowing river for three minutes while kicking the river bed with the other foot and holding on to a net that the river is trying to sweep away ensures that these guys don't need to go to the gym. I demonstrated this technique and offered my best health and safety advice; then the children donned their wellies and waded into battle with the mighty River Walkham. The morning was filled with flailing arms and nets as apprentice aquatic ecologists desperately tried to keep their balance on the rocky river bed. A conveyor belt of excited, net-bearing children moved from the river to the bank with nets full of caddisfly tubes and stonefly larvae that were deposited into the waiting white trays. There was much poking and oohing as caddis larvae slipped their heads and legs from their tubes and dragged their bodies around the tray. Diving beetles shot around the trays like sleek black torpedoes, while damsel- and dragonfly larvae walked majestically along the bottom of the trays daring anything to approach. It was not just the nets that were full but the children's wellies were also overflowing with samples of the Walkham river as caution was thrown to the wind in this maelstrom of aquatic excitement. By the end of the morning, we had twenty soaking wet children who were emptying their Wellington boots onto the riverbank. An air of calm then descended on the riverbank as the children's energy waned; their voices were quieter, but laughter still trickled along the riverbank. The river may have exhausted them physically, but the morning had left their curiosity in overdrive. They were now soaked to the skin, tired and hungry but had seen many strange and curious animals that they had never encountered before.

These events became a regular part of my summer. To help the children identify what we found, I produced a sheet with a flowchart that guided them through the various groups of invertebrates. This was very basic and over the next two years evolved into a two-page guide dramatically labeled *A Guide to the Bugs and other small monsters of Tavistock*. Word of my willingness to take small children bug hunting had spread and Jenny Hale, a local potter and wildlife artist, invited me to run a bug hunt for her Wildlife Watch Group at Brent or on Dartmoor. While there I met John Walters, a local artist and entomologist who thought that my sheet had potential and could be expanded into a booklet. After a couple of meetings, we set forth to do so and, finally, after ten years of testing various versions on successive groups of children, we produced *Minibeasts, a guide to invertebrates*. I wrote the identification guide, while John added the drawings of all the minibeasts. The book aimed to allow children to put a name to any invertebrate that they might find in a school field or their garden. It was meant for young people but also catered for adults who wanted a basic introduction to the topic. The draft was well received by the many entomologists we sent it to for review and was then revised in light of their comments. Dick Vane-Wright, who was then Keeper of Entomology at the Natural History Museum, wrote a foreword and the presidents of both the Royal Entomological Society and Amateur Entomology Society wrote an endorsement. It sold well and most of the 6,000 copies we had printed have been sold. John and I sometimes talk of an updated second edition; it's on our list of future projects.

In the mid-90s, the deputy Vice Chancellor of the then Polytechnic South West received a huge research grant from the Biotechnology and Biological Sciences Research Council (BBSRC) that came with the caveat that he had to spend some of the money on Outreach activities. He was not too keen on the idea of going into local schools, so he passed this responsibility on to my colleague Karen Gresty, who then took on the task of coordinating this outreach and the associated money. As she was aware of my weekend activities with the local Wildlife Watch Groups, I was drawn into the frame to help make it happen. Working with Karen, we visited local schools to give talks and ran bug hunts, plus we also developed a quiz using the more unusual specimens from the

Students from Tavistock College wrestling with the Natural History Quiz.

department's museum that imaginatively became known as the Natural History Quiz.

The money from the grant soon ran out, but shortly after we had set up our Outreach group, the BBSRC decided to form a national schools' Outreach forum to encourage universities and research establishments to engage with their local schools. Each group would receive an annual grant of £800 to spend on linking with local schools. We were back in business to run school visits and activities which we did for the next ten years. By this time Outreach became officially recognised by senior management and full-time staff were appointed to run the show, but I continued to engage with local schools under the official radar.

Several activities came out of this, one of which was 'Build a Bug'. The idea was to introduce children to the major groups of invertebrates that we have in the UK and get them to build a model that had the correct number of body parts and legs, plus the correct number and shape of wings. These models were built

School children build bugs at Waterstones in
Plymouth as part of the National Insect Week.

from coffee cups and drinking straws and were held together with PVC electrical tape (this tape was great as it came in a variety of colours and was cheap to buy). I also produced faces for each of the invertebrate groups to give them character. The children loved this mix of science and craft, proudly taking their mini-beasts home to terrify younger siblings or parents. While butterflies and bees were always popular, millipedes and centipedes were sometimes seized upon. I recall one school where the children made a model centipede that took ten children to carry it. This monster was then proudly borne from the classroom through the school to the headmaster's office. Build a Bug is alive and well and still makes regular appearances at the York and Bristol Insect Festivals and other events around the country.

Weird Worlds

In 2002, Plymouth City Museum approached the Faculty of Science at Plymouth University offering them their main gallery for eight weeks to put something on. The Dean felt this was an excellent way of promoting what we do in the faculty and called a meeting to explore possibilities. He had decided that each of the six departments would have a week in which to showcase the department's activities. Consequently, department heads were dispatched to dream up ideas for a week-long exhibition. This was a considerable challenge for busy academics, hence the Biology Department put out a call to all staff for ideas. I had always been fascinated by novels that explored a 'what if' scenario – *Pavane* by Keith Roberts is a great example. What if Queen Elizabeth I had been assassinated? How would the UK look today? It's a very scary vision, but it's an intriguing book.

I decided to apply this concept to entomology and considered what would the world look like without various groups of invertebrates? Woodlice consume a vast volume of dead leaves each year … what might an English woodland be like without them? This was also a chance to make the public aware of some of the very useful things that invertebrates do. I decided on woodlice, dung beetles, ants and spiders. The display would feature panels eight feet tall, each one offering a view of the world without a particular invertebrate plus information panels which described their biology. The ant panel would house a colony of leafcutter ants as I had seen these at London Zoo where they were very popular with the public. The final panel would offer a global view of invertebrate diversity with lists of the numbers of insects in various groups plus a revolving globe. I submitted my proposal and waited for the cries of 'You are joking! This will cost a fortune!' To my surprise I was given the green light and told to get on with it. My colleagues in microbiology had offered a series of activities dealing with bacteria, so our week was labeled 'Bugs', and would be an exploration of both colloquial meanings of the word.

Panic then set in! What had I been thinking? Who would design these panels? Who would build them and where did one acquire a colony of leafcutter ants? There was also the question of when would I be doing this as I already had a job that took up all of my working day?

The Dean was great; he was totally supportive and seemed to throw money at the project. I approached the university carpenter with designs for the panels; he had already done a wonderful job with the set for *Real Bugs* (see chapter 8), and after a few modifications set to work. Our graphic designers were equally helpful, coming up with novel ideas to enhance the panels. We soon had our panels drawn out in rough. An introductory panel that explained why invertebrates are useful, whilst the woodlice panel showed a woodland with just the tree top protruding above the dead leaves. The dung beetle panel showed a mountain of rabbit droppings, and the spider panel presented a spider in a web with a pile of aphids filling the bottom third of the panel. The ant panel had a gaping hole which awaited the ant colony; the globe for the final panel was still to be made … but it was coming together. We had regular meetings with Helen Fothergill, the curator of Natural History at the museum, and it all seemed to be going well. The ants were still a problem. I had seen leafcutter ants at London Zoo, but where to obtain a colony? I remembered that Paignton Zoo used to have a colony, so I approached their entomologist, David Stradling, a man that I knew had worked on leafcutter ants in the past. David was delighted to be of help as supplying and setting up leafcutter ant colonies had become his hobby in retirement.

It seemed that he had cornered the market in the provision of leafcutter ant colonies; he now made regular trips to the Caribbean where these ants are considered a pest, and so the authorities there were not concerned if he took a few colonies back to the UK. When I visited his home, he had four colonies set up in his dining room awaiting dispatch to their new locations. I did not want to buy a colony, just borrow it for a month. A price was agreed and a date set for delivery.

That was all very well, but what to use to put it in? I would need a large glass tank so I could set up two islands inside: one with the

The Weird Worlds exhibition panels.

ant colony and one where the leaves were placed. These would be connected by a few twigs. David had warned me of condensation on the glass and recommended double glazing. I approached a local glazing company who made a double-glazed front panel and then cut the glass for the rest of the tank. My colleagues and I then had to glue them all together with silicone.

The day finally arrived when the panels were built, the graphics printed and glued to them. A football had been transformed into a globe and a small motor installed to keep the world turning. Everything was set up. The tank was in place behind the hole in the ant panel, and our carpenter had built end panels to prevent the public from venturing behind the scenes. We awaited the ants. David arrived and to my surprise he had the ant colony in an open top cardboard box. "Oh well," I thought. "He is here now, what can go wrong?"

We approached the gallery where the exhibition was set up. The doors opened both ways to speed the entrance and exit of the public. When we arrived, Helen Fothergill, the curator, was just leaving the gallery. David and I were in deep conversation and Helen was on an urgent mission. The doors swung towards us as Helen hurried out, just catching the box of ants! David desperately tried to retain his hold on them, but they spiralled out of his hands through the now open door and landed in the gallery – exploding leafcutter ants across the floor!

Time slowed to a halt. Ants tumbled and slid across the parquet flooring. We were rigid with horror; fear and dread were stamped on each of our faces. David and I were terrified that we had lost the colony and Helen was terrified that she had just lost her job. We went pale at the thought of hundreds of leafcutter ants rampaging through the museum collections. Both our exhibition and Helen's future career hung in the balance. Everyone in the gallery rushed to our aid. Anything that was to hand was brought into play to capture the dispersing ants. Sheets of paper, paper tissues and the odd envelope were employed to sweep the ants back into their container. After twenty minutes of feverish activity the last ant was cornered, and the day was saved. The exhibition could now proceed and Helen's future at the museum was likewise secured. The ants were installed, and they quickly found their supply of leaves. The exhibition was now set: we would have lunchtime lectures on entomology and to encourage schools to visit laid on the *Real Bugs* puppet show each afternoon (see chapter 8).

The exhibition was a great success with thousands of people coming through the doors. One fascinating piece of work to come out of it was a survey conducted by the Psychology Department at Plymouth University who examined children's perception of scientists. Staff from the department would visit local schools before their visit to the exhibitions and asked the children to draw what they thought a scientist would look like. Most of these drawings depicted an older male with receding or frizzy hair, glasses and a general wild appearance. The children perceived them as unfriendly and even dangerous. Would any of them want to be a scientist when they grew up? Decidedly not.

Following their visit to the exhibition, the psychologists returned to ask the same questions and requested a second drawing. This time scientists were young people of both sexes who were friendly and keen to make the world a better place. Many of the children now thought science might be an interesting area to study. QED, scientists are just ordinary people and not the monsters portrayed by Hollywood.

As my role in the department's Outreach programme had grown, I decided that it would be a good idea to include an Outreach session at the Royal Entomological Society's annual conference. I floated the idea and received a positive response from the convenors. The next meeting would be at Reading University and an Outreach session was now on the programme. I decided to talk about the sessions we had run at Plymouth and had also asked Kieren Pitts from the Bug Club to give a talk about its activities. In addition, I was keen to have something of the role that was played on TV, thereby raising public awareness of insects and the wider natural world, especially after talking to a group of first-year marine biologists. I had asked them what had persuaded them to study marine biology and almost all of them said they had been inspired by the BBC series *Blue Planet*. It appeared that television was a very persuasive force. So I phoned the BBC in Bristol to ask if anyone might be available and willing to talk about television and its portrayal of insects. After a few false starts of "Ah, no, that's not me. Have you tried this number?", I was eventually put through to Simon Williams. I asked if he or any of his colleagues might be willing to give a talk at the Royal Entomological Society's meeting on 'Insects in television programmes'. I was met with a long pause.

"How did you know?" he said.

"How did I know what?" I replied.

"About the series."

"What series?"

"So, you don't know then."

"No, I don't know. Tell me about it."

It transpired that he had been in this job for just a week. He was part of a small team of researchers who had been employed to gather stories for a new series on insects that would be called *Life in the Undergrowth* and presented by Sir David Attenborough, another of Sir David's famous *Life* series. I had inadvertently struck gold. Simon and his colleague came and gave an excellent talk on the way that TV portrayed insects and the new high-tech methods they now had at their disposal plus just enough about *Life in the Undergrowth* to pique our interest in the coming series. They were, of course, also keen to network and track down new stories for the series.

Having made this connection, Simon's team often phoned asking for contacts regarding entomological stories, and over the following three years I managed to connect them with several projects.

As the series approached completion, I asked if someone could offer a talk at that year's annual conference to be held in Brighton. Mike Salisbury, the series producer, offered to come along. Interest in Outreach had increased over the previous three years. We now had five talks in the session, with Mike's talk being a blazing comet compared to the rest of us. He explained the complexity of filming the series, talked about some of the biologists he had worked with and showed a couple of clips. The session was well received, and subsequently Outreach sessions have now become regular additions to the Royal Entomological Society's conference programme.

I was not the only member of the Society to be involved in researching *Life in the Undergrowth.* When the programme's launch party was announced at the BAFTA headquarters in London, about ten of us were very excited to receive invitations to the event. But this was not a conventional invitation; it was a scroll of paper eighteen inches long bearing the words 'Life in the Undergrowth' with each letter made up from a horde of invertebrates. The only other mark on the scroll was the BBC logo. The practical details of where and when were sent by e-mail.

Invitation to the *Life in the Undergrowth* launch.

I had recently taken on the role of co-editor for the Royal Entomological Society's house magazine *Antenna*, and it just so happened that the editor, Greg Masters, was in the process of transforming it from an A5, mainly black and white publication, to a glossy full colour A4 magazine. On the day of the launch, we had a meeting at the Society's headquarters in South Kensington, London, to discuss the new format for the magazine further and decide on the content of the first edition. We were looking for a dramatic story to launch the new look of *Antenna*. I recall Greg joking: "Why don't we ask Sir David to give us an interview about his new series? That would really launch the new magazine." Wow, what a scoop that would be, but seriously, how likely was that? Our colleagues at the meeting, however, thought it a great idea and our smiles soon faded as we realised that they were serious. We travelled across London to the *Life in the Undergrowth* launch with a certain pressure on our shoulders. Could we pull this off?

At the launch we were presented with some extremely exotic canapés while mingling with colleagues and the media. The proceedings began with a talk by Mike Salisbury, followed by a Q&A with Mike and Sir David. We were treated to a brief excerpt of the programme with more drinks and mingling to follow. Sir David was constantly surrounded by a tight group of people; there was little chance of gaining his attention and, anyway, what was I going to say. What could I say to my boyhood hero that he had not heard a thousand times before? Success looked like a distant prospect.

Di Burton, the Royal Entomological Society's then PR agent, appeared at my elbow asking, "Would you like to meet Sir David?" I wanted to say "No, he looks far too busy", but heard myself say "Yes, of course!" Di grabbed my arm and led me across the room. My courage began to fail me as we approached the tight knot of people around the main man. She tapped Sir David lightly on the shoulder and said, "Please, excuse me Sir David, but can I introduce Peter Smithers from the Royal Entomological Society?" He turned and smiled. I thrust my hand forward. "It's a great honour, Sir David. This is the finest series you have yet produced. *Life in the Undergrowth* is amazing." He smiled and gripped my hand. "Why, thank you," he said. There was a pause. The opportunity was slipping away. I had to strike now. "Sir David, the Royal Entomological Society is just revamping its magazine and we wondered if you might grant us an interview to launch the first edition?" There was no pause, no hesitation. A simple "Yes, of course, here is my card, drop me a line and arrange a time. Very pleased to have met you." He was whisked away and I was left standing there with my colleague at my elbow and his card in my hand. No one could believe it was that simple; I was expecting agents to be involved and delicate negotiations … but no, just his card.

A letter was then written and a date arranged for the interview. We decided that we would meet him at the Royal Entomological Society's HQ, then at 41 Queen's Gate, South Kensington. We would begin the interview there and then take him to lunch in a nearby hotel. The interview would be conducted by Greg Masters and I as editors of *Antenna*, plus Bill Blakemore, the Society's CEO. Meeting one's heroes is always a daunting prospect and so it was with a mixture of apprehension and excitement that I walked along Queen's Gate on a cold February morning with sleet driving out of the wind. Greg was already there; we awaited our guest's arrival in Bill's office. A tense expectant atmosphere descended upon us.

The security cameras revealed his approach to the front door of 41, Queen's Gate and Bill's PA came in to announce "Sir David has arrived. Shall I show him in?" We stood in unison. I recall thinking "What can I say to a man who has been greeted so many times?" I settled for "Good morning."

Bill introduced everyone and organised coffee. Sir David surveyed the three of us, pulling a tie from his pocket. "I have brought a tie in case we are going somewhere posh for lunch." He looked around. Bill and Greg were both sporting the Society's tie. He turned to me. "He's not wearing one, so I am sure I won't need it!", putting the tie back in his jacket. This informality broke the spell and we all relaxed. He then told us that he had been visiting the rare book collection at Windsor Castle in preparation for a new book, *Rare Beautiful Things*. He was disappointed that they did not have *The Aurelian or Natural History of English Insects, namely, Moths and Butterflies, together with the plants on which they feed* (1766) by Moses Harris, the first book published on British butterflies. "Ah ha," said Bill. "We have a copy in our vault. Would you like to see it?" Of course, he did.

Bill brought it out with a selection of the other rare books that are kept in the Society's vault. Wow, we thought. Showing Sir David something he had not seen before must be a rare event indeed. We perused *The Aurelian* and Maria Sibylla Merian's 1705 *Metamorphosis Insectorum Surinamensium*, the first book to show the complete lifecycle of butterflies, including their food plants. Having impressed Sir David, we settled down over coffee to discuss his life and the success of his new TV series *Life in the Undergrowth*.

> PS: "This is the first time you have produced an entire series that looks at insects. Are you an entomologist at heart?"
>
> DA: "I am not an entomologist as I am completely incapable of making the study of one group of organisms my life's work. There are too many other interesting animals and plants to investigate, and I prefer to take a more holistic view of the biosphere. The insects are, of course, extremely important and exhibit a huge range of fascinating behaviours and life

histories. So it's impossible to ignore them when making programmes about the natural world. As a result, almost all of my programmes feature insects in some guise."

PS: "How did *Life in the Undergrowth* come about?"

DA: "The idea of producing a programme about invertebrates had been at the back of my mind for some time but getting decent close-up footage had been a problem. However, with the recent development of cameras that were small and had the enormous depth of field required to film insects, it was now a realistic proposition. With this in mind, I had begun to sketch out a rough story board for five programmes.

At a meeting with the BBC programme controller, I was asked if I had any ideas for future programmes. This was my cue, so I launched in with my outline of five programmes on invertebrates.

'Fine', he said, 'go ahead.'

On my return home I quickly drew up plans for a couple of extra programmes hoping I could extend the series and made contact with the controller the next day but was told 'No, five is fine, go with that.'

Bother, I should have asked for more initially.

I had wanted to start with the invasion of the land and then move on to the evolution of flight. A programme on spiders was a crucial part of the story, but I felt that as they were the least popular group to start the series with, it may put people off. So we decided to begin with silk and its use in the insect world and then move sideways into the spider's realm. The last programmes would then explore interrelationships and insect societies."

Greg and I were curious that with such an ambitious series there must be stories that did not make it into the programmes.

DA: "In fact there were very few. The planning process tends to preclude too much unwanted footage. Once the story boards had been finalised and the location and time of the main elements had been fixed, we then look to see if there are other stories that can be filmed at nearby locations. By fixing our locations and times to get specific behaviours our schedule fills up rapidly and we then have to drop other stories as we just can't get to them in time."

As entomologists we felt that the series had been a great success, so we asked Sir David how he felt it had turned out:

DA: "On the way here the cab driver recognised me and was very enthusiastic about the programme. 'Bloody marvelous, Sir David', he said. 'I had no idea how fascinating the little blighters could be.' In fact, the series was very well received by the BBC as well. They assess programmes in two ways: the number of people who watch a programme and the Audience Appreciation Index. This is an online survey of 20,000 volunteers that generates a score between 1 and 100. The first *Life in the Undergrowth* programme had a score of 91, the highest score for any BBC programme of its kind ever broadcast." (At that time.)

Lunch was calling and we walked to the Strathmore Hotel where we had booked a table in their restaurant. The streets were not busy, but I was extremely conscious that I was walking, deep in conversation with one of the best-known faces in the UK. Heads turned briefly as we passed by, but this was London and celebrity faces are commonplace on these pavements. As we walked along, I asked him what he had planned next.

DA: "I am currently working on a series about reptiles and amphibians, *Life in Cold Blood*, which should be on your screens in two years' time. There is also a programme on climate change, but as the current picture keeps changing, we are hard pushed to keep pace. It is bound to be out of date before it is screened."

We arrived at the hotel and were shown into the bar. I asked Sir David what he would like to drink.

"Just a mineral water, please."

Then he looked at me.

"What are you having?"
"I am having a beer," I replied.
"Then I will have one too."

Lunch arrived as we were discussing the state of natural history in UK education.

DA: "The lack of natural history in schools is awful, but there are many reasons for this drift away from natural sciences. When I was a boy, I collected a wide range of natural objects which I kept in my room. These days, children are discouraged from collecting anything, often for very good reasons, but an interest in the natural world comes from observing it at close quarters and without a personal museum this is much harder to develop. Children are no longer allowed to wander the countryside as I did, and more people now live in towns, so there is less opportunity to get into the countryside. Combined with this is a greater range of distractions and activities for young people; it is not surprising that fewer young people get involved in the natural sciences. However, I do find that the interest is always there, one just has to trigger and harness it."

Sentiments that we all agreed on. As a member of the Outreach team at Plymouth University, I was constantly aware of the threat. The only natural history primary school children around Plymouth received was when I went in to deliver it.

Lunch had come to an end and we were on our second coffee. There were too many questions and only so much time, but I had to ask one last question.

PS: "After five decades of making natural history programmes, what does the future hold?"

DA: "I feel that once *Life in Cold Blood* is finished, I will have covered all of the major groups of animals and plants, plus all the major habitats (*Living Planet* and *Blue Planet*), and behaviours and life histories (*Trials of Life*). I hope that the *Life* series will stand as a record of the state of the natural world at the end of the 20th Century and will provide a reference point for natural historians in the future. What happens after *Life in Cold Blood* I am not sure. I have just turned 80, so I am not so young. I can't climb trees anymore. If someone in their 80s had approached me when I was controller of BBC 2 asking to make a new series, I would have been extremely hesitant. When you ask the BBC for money, the controllers fix you with that steely eye and want to know what they will get in return. We will have to wait and see what I can offer.

On that note, I must be off. I have a few items to return to the Natural History Museum."

We shook hands and he departed.[4*]

Meeting my boyhood hero had been wonderful. He was unassuming and generous with his praise of others. I am not surprised that he had been recently voted 'The Most Trustworthy Man in Britain'. He remains an adventurer hunting for those stories that reveal the wonders of the natural world, a man dedicated to sharing these stories with the rest of us, a man who still has plenty to offer. As the decades since this interview have shown, he

---

4   Sections of this interview were first published in *Antenna* Vol. 30 (2) and are reproduced here with the kind permission of the Royal Entomological Society.

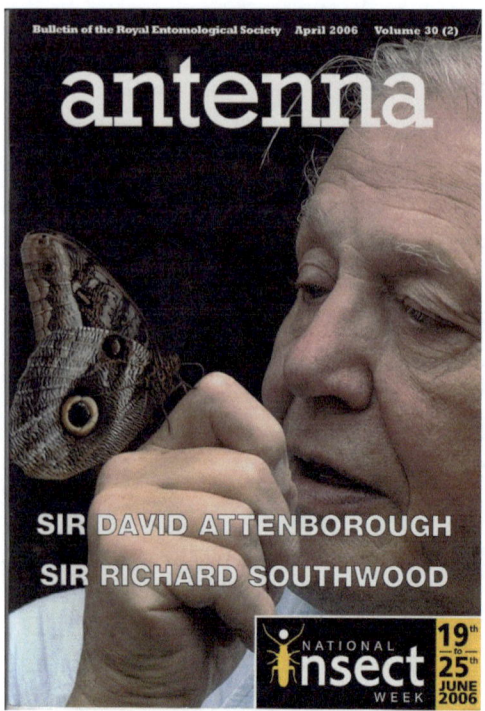

The cover of the first A4 full-colour edition of *Antenna*.

continues to narrate all the major natural history series that the BBC produces.

The interview launched the new *Antenna*, and it has never looked back. Over the years it has evolved into a magazine that reflects the diverse nature of entomology, exploring the core and frontier aspects of entomological science, the links with modern culture, entomological history and the entomologists themselves. I edited the magazine for another twelve years before handing the reins over to my younger colleagues.

Investigating insects

Having an inside track on the BBC's new TV series *Life in the Undergrowth*, I decided to plan ahead as I was certain that this would generate a wave of public interest in insects. The Weird Worlds exhibition had been an outstanding success, hence Helen Fothergill and I decided to plan an exhibition about insects to coincide with the screening of the series. This would comprise details of local entomologists, both past and present. A series of scanning electron micrographs of insect heads, rare entomological books from the museum library, and scans of images of other rare books from the Royal Entomological Society library. There were also several insect illustrations by local artists.

The exhibition came together and was to be part of a launch weekend for *Life in the Undergrowth*. Whilst at the launch party in London I persuaded the series producer, Mike Salisbury, to come to Plymouth and give a public talk in the Biology Department about the making of the programme. I would follow this with a talk on insects as food and we would then provide some insect snacks and lead people over the road to the exhibition in the museum and a performance of the puppet show Real Bugs (see chapter 8 on Insects and Art).

Poster for the exhibition Investigating Insects

This preview session was held the day before the first programme was to be screened, so there was a great deal of interest. The 110-seater lecture theatre was packed, with people standing at the back and others sitting in the isles (something that I would never get away with today). Mike Salisbury could not make it, but he had sent along his right-hand man, Peter Basset, who had produced the first programme. Peter gave a wonderful talk that revealed how difficult the programme had been to make and the surprisingly vast volume of equipment that they had taken with them. He also showed several clips from the first programme to illustrate his points, which delighted the audience.

Room layout for Investigating Insects

There was a short break and then I gave my talk on insects as food to an almost packed house. I did not have quite the same attraction as the BBC. In the break, I had set out the edible insects on tables outside the lecture theatre and several students were on hand to answer questions and hand out the snacks. To my amazement the queue went around the foyer and out of the main door. People were curious – if not hungry – and once again everything was eaten.

Helen Fothergill had laid a trail of beetle stickers on the pavement from the university to the museum. After nibbling a few insects, many of the families hit the beetle trail and headed for the museum and a performance of the puppet play *Real Bugs* which left plenty of time to look around the exhibition.

Over the next month the museum ran a series of public lectures that linked to the exhibition, and I had arranged an entomological theme to the department's autumn lunchtime lecture series. The Biology Department at Plymouth University had a truly very entomologically biased autumn that year.

Workshop under the sky

Jennie Hale is a potter, wildlife artist and natural historian, a lady of vibrant enthusiasm and boundless energy. When I first met her, she lived on a remote farm on the western edge of Dartmoor and in true Jennie style made her ceramic animals and pots using the Raku method. This is a finishing process that is as wild, dramatic and untamed as the potter herself. Glazed items come out of the kiln and are lowered into a crate of wood shavings. These catch fire and a storm of sparks and flames roar into the air. As the pots cool rapidly, the glaze crazes and the burning wood is deposited in the cracks forming a black network across the surface. This adds a wonderful extra dimension to her work; her pots and ceramic animals are highly prized. My wife has one of Jennie's foxes that sits alert and quizzical in our hall, assessing our visitors with its mischievous eyes.

Her illustrated wildlife diaries had won the BBC Wildlife Magazine's competition the first time it ran and have since been published as *The Nature Diary of an Artist*. Her love of art and wildlife was the main driving force in her life. Following the success of her 'Wildlife Watch' group, Jennie wanted a more creative approach to engaging young minds with the natural world. The 'Workshop Under the Sky' was born as a collaboration between scientists and artists working together to bring the magic and wonder of the natural world to children in the countryside. 'Workshop Under the Sky' is based in a yurt, a wooden-framed tent, which is erected on the site to be investigated. Scientists take children around the local habitats explaining the lives of the plants, insects and other animals that are found there. Then the children are taken back to the yurt where artists work with them to draw the wildlife they had found. These were wonderful days, full of laughter and curiosity. During the second-ever National Insect Week, Jennie worked with the Forestry Commission to use Lydford Forest, on the edge of Dartmoor, as a base. The local ranger was on board, so the yurt was set up by a pond in the heart of the forest.

Moth traps were run each night and, in the morning, as the children arrived, we would show them the catch. It was magical: Jennie and her friend Terry (the chief yurt erector) camped in the yurt for the week, and I traveled out from Plymouth each day. I often took a couple of students with me and we were joined by local entomologist Kevin Brown. These mornings were full of unbridled energy as the children raced down the forest rides sweeping nets through the taller vegetation and holding umbrellas under branches while they tapped them with bamboo canes to dislodge any resident invertebrates. We had also set a series of pitfall traps to catch the beetles or spiders that walked around the forest at night. A vast array of beetles, bugs, millipedes, woodlice, spiders and assorted other invertebrates were flushed out of the vegetation and scrutinised. Kevin, my students and I would do our best to talk about them, but the questions from the children then came thick and fast. A veritable tsunami of demands for knowledge driven by their unfettered curiosity.

The call for lunch would calm the mood to one of excited chatter and the children would adjourn to the yurt to draw selected plants and invertebrates that they had collected, aided and encouraged by Jennie's team of artists. A quiet and intense concentration pervaded the afternoon. As each drawing was completed, it was hung in the yurt and another was begun. Before long, the inside of the yurt was covered with drawings and paintings of local insects, spiders, millipedes and plants. Some of these drawings were stunning and some imaginative, but they all spoke of a period of intense observation and an appreciation of the plants and animals that the children had seen. At the end of the week Jennie made an A2-size book of the best drawings from each school and presented a copy of this collection to all the schools that had attended. I still have my copy of the Lydford Forest book, and it still wows friends and colleagues who browse through its pages. When I retired, I moved to Bristol and dropped out of the team, but the 'Workshop Under the Sky' is still running today, continuing to inspire and engage curious young minds with the natural world.

Biotropica

During my time at Plymouth, I organised a series of lunchtime lectures for students entitled Exploration Biology. These were part travel talk and part research. The speaker would describe the work they had gone to do, talk about the local problems such as equipment failures, how good or bad the food had been, encounters with local wildlife and local people, and occasionally how bad the toilets were.

These were a taster to introduce the students to the realities of biological research in the field. Many of these lectures were given by staff in the Biology Department, but students who had volunteered for overseas conservation projects were another rich seam of volunteers and I was soon inviting anyone

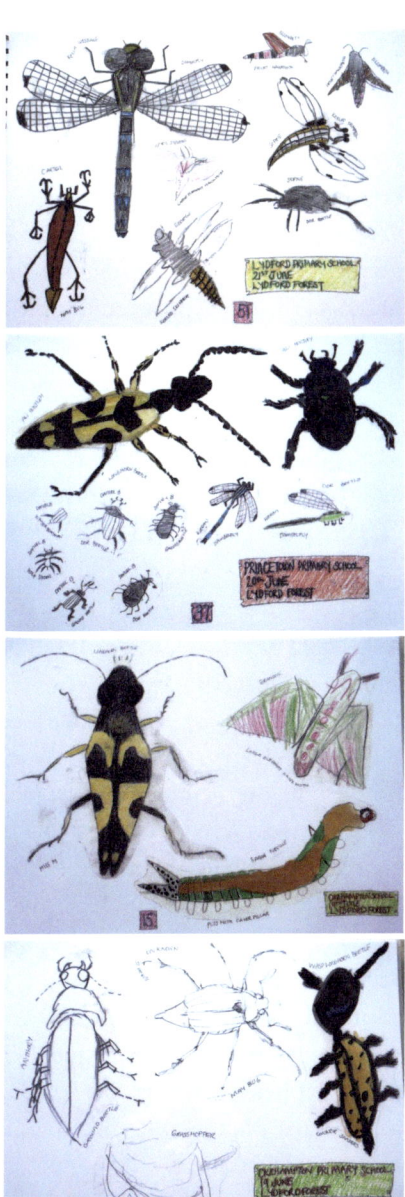

Pages from *A Forest Full of Bugs*. Jennie Hale's compilation of children's drawings made at Workshop Under the Sky.

I knew who had been somewhere interesting to give a talk. Topics ranged from saving turtles in the Caribbean and conserving their offspring to exterminating rats on islands in order to conserve local wildlife. From visiting hydrothermal vents in a submersible and exploring rainforest canopies to the biology of the land crabs on Christmas Island. I can still recall Antony Jinman's talk on polar exploration in which he described the frantic process of taking a photograph under polar conditions. The camera batteries had to be kept next to the skin to keep them warm enough to maintain their charge. Therefore, clothes had to be undone in order to reach the batteries, gloves came off to handle them and all at subzero temperatures. Photography was a race against the onset of frostbite and battery failure. More personal activities were just a nightmare. Despite this, Antony had some amazing photographs to share with us. These talks were always popular, despite the students having to rush on to the next lecture via a quick sandwich. Many of them were inspired to go on to take up the challenge of research in remote corners of the earth.

While organising these talks I became aware that the staff in the Biology Department had a vast archive of images that were languishing on their hard drives, making only brief appearances in their lectures. Why not collect the best of them together to make an exhibition that would further inspire our students. I knew the head of the Photography Department, Liz Nicol, so I asked her if she would help with this project, and she was very keen to do so. I approached my colleagues asking them to offer six photos from their collection and then invited Andy Foggo, Associate Professor of Ecology, to be one of the judges as I knew he was an excellent photographer. Liz would also be a judge along with Jem Southam, a landscape photographer of international repute who was also a member of her group.

A call went out, I twisted a few arms and we had over hundred photographs to sort through. This was my first experience of the very different perspectives that artists and scientists have of the world around them. Andy and I would look at a picture and wax lyrical over its precision, focus and biological content, while Liz and Jem would stare in amazement claiming that the said image had no merit at all due to its dull predictability. It would make an ideal chocolate box cover! At other times Liz and Jen would become very excited over an image that Andy and I felt was mundane and dull, but eventually we agreed to compromise and slowly, ever so slowly, an excellent series of photographs were selected for the exhibition. The images hung in the university's Cube Gallery for two months before being dispersed around the biology department where they still reside.

Biotropica poster: Wildebeests at dusk. Photo by Andy Foggo.

A cicada emerging from its nymphal case. Photo by Rob Parkinson.

Rick drawing a St Andrews cross spider,
(*Argiope* species). Photo by Peter Smithers.

Michelin stars and celebrities

Now that I had a reputation as the man who eats insects (see chapter 6, A New Hope) I receive occasional requests to attend events and persuade people to try a few crunchy insect snacks. One such event was at Plymouth Pavilions, a large entertainment venue for concerts and conferences. They had decided to branch out and put on a leisure fair that brought together many of the region's outdoor activity businesses to showcase the wide range of adventure activities available in the region. In order to raise the profile of this event, the organisers had engaged bush craft expert Ray Mears to open the day. As Ray is well known for his survival programmes, they approached me asking if I would man a stand offering insect snacks to try. They envisaged a wonderful photograph of Ray eating insects with local school children to raise the profile of the event in the local press.

No problem from my perspective, so I began to plan the range of snacks I would offer. Then the university's PR team approached me asking if I would be happy to work with a local chef, Peter Gorton, who had been going into local schools to inspire children about food and cooking. I was intrigued at the idea of working with a proper chef and agreed. A meeting was set up. Peter and I gelled immediately. I talked about my fascination with insects as food and Peter talked about his work with local schools. He was keen to interact with people at the event. I offered to provide a container of crickets and Peter agreed to turn them into an interesting dish. This was brilliant; I could offer the plain snacks in the morning and a freshly cooked dish prepared by a Michelin star chef at lunch time.

Wow, a Michelin star chef would be cooking insects with me; how cool would that be! I turned up with my wares and was shown to a table that would be my stand. I laid out the insect snacks along with their labels and placed the cool box of crickets under the table to keep it safe. The organisers of the event dropped by to make sure all was well and introduced me to a group of children who would be posing for the photograph with Ray Mears. They eyed up my insect snacks suspiciously, so I ate a couple of crickets to reassure them. They didn't look convinced.

The organisers slipped away to meet and greet their celebrity, leaving me to wait with the children. The team returned with Ray Mears in tow. Introductions were made and the concept of the promotional photograph was explained. I stepped in and explained what the dishes on offer were. The organisers were nodding, but Mears was looking uncertain. He looked at the insects on offer and simply said: "I am not doing this. It's just sensationalism. More of *I am a celebrity, get me out of here* rubbish. No, I refuse to pose for this photograph."

Silence. The organisers were looking at each other in bewilderment. They couldn't believe this was happening. They huddled together in whispered conversation and then drifted away to find a solution, leaving Mears and me standing next to a table full of insects he didn't want to eat. There was a further awkward silence.

Ray turned to me and said, "I really disapprove of this celebrity-get-me-out-of-here approach; it gives the wrong message to the public." "I heartily agree," I replied. "But that's not why I am here!" I launched into the reasons why insects might make an important contribution to feeding a growing world population. The barriers came down and we were suddenly both on the same page. Ray opened up with stories of insect eating he had encountered on his travels and, while he wouldn't eat any insects, he said he would pose for the picture with the school children. The organisers breathed a sigh of relief; the photos were taken and Ray Mears was whisked away to see the other stands. I offered my insect snacks for an hour, then Peter Gorton arrived. Wow, he was enthusiastic – a whirlwind of culinary energy. He was set up and cooking in minutes, preparing a vegetable risotto to which he would later add the crickets.

The crowd was already gathering, and we both took the chance to talk to them about what we were going to do. We bounced off each other telling them why insects are an important food and why cooking them is important. The risotto was served – and gone in a flash. I couldn't believe it. We had served Micheline-star cricket risotto and it went in moments. We had also had lots of fun; a partnership had been born.

Bugs, film festival

Film has always been a passion of mine. I have been extremely envious of the Insect Fear Film Festival that May Berenbaum runs annually at the University of Illinois in the USA. She would show a couple of insect horror movies over an evening with a short introduction by an entomologist to give the audience an informed perspective. There were also lots of entomological-based stands in the foyer outside the cinema to entertain and educate the viewers before they saw the movies. I had often thought about replicating this but a venue was always the problem. However, when Plymouth University gained a new arts building which included a state-of-the-art cinema, I saw possibilities and explored them.

I approached David McLaren who had been appointed to run the cinema and discussed the idea with him. David was wildly enthusiastic. The Plymouth Insect Fear Film Festival was on. We outlined a plan for the day: children's films in the afternoon, such as *A Bug's Life*, along with a selection of short films, such as *Smalltalk Diaries*, which overlay a humorous voice on films of real insects, and *Minuscule*, a French animation that is very funny. At the end of the afternoon, we would screen '*Them*', the 50s film about giant ants that appeared in the Nevada desert after the atom bomb tests. For the evening screening David had selected *Bug*, a film about the fear of insects. He was so enthusiastic about this film despite never having seen it. I agreed. "You are going to love it," he said.

We organised a range of stands for the foyer outside the cinema. Jan Freedman from the City Museum agreed to have cases of insects on display from their collections; Buglife and Butterfly Conservation were also there offering information on insect conservation, along with a team from the Devon Wildlife Trust. Dartmoor Zoo were joining us with a range of livestock, both

Poster for the BUGS 2 insect film festival

invertebrate and vertebrate, and I had organised students from the Biology Department to offer face painting, build a bug and insect masks to make and take home. The foyer would be bustling. I made contact with Peter Gorton to see if he might be interested in coming along in the afternoon to cook an insect-based meal in-between films. To my delight he was keen to work with me again and agreed to come along.

We still needed a big name to draw the audience in. I contacted the TV presenter and naturalist Nick Baker. I already knew Nick from an interview that Greg Masters and I had done for the Royal Entomological Society's magazine *Antenna*, so it was a very informal approach; an email was followed by a phone call. Yes, he said he would be delighted.

The day arrived. The tables were set out in the foyer, all the films were stacked in the projection room and stall holders were setting up shop. Families began to appear and visit the various stalls. Nick arrived in a shirt dotted with ants and chatted with families in the foyer. All was going well. We moved into the cinema and Nick gave his usual wildly enthusiastic talk on how he became interested in insects and then introduced the first film, *A Bug's Life*. The audience were having a great time. Nick stayed for the rest of the afternoon taking in the movies and chatting with everyone.

Peter Gorton arrived to set up his stall, which was now known as the *Ento Café*. This happened at a lightening pace, with a hot plate and an oven ready to go in moments. A crowd gathered and we launched into preparing food. Peter had brought the ingredients to make a mango and cricket risotto. There was much excitement. The rice, veg and fruits were prepared and cooked and, with a dramatic flourish, in went the tub of black crickets. There was a palpable hush in the audience at this point. Completed, the risotto was handed out in plastic cups with a teaspoon. Initially people were tentatively trying the dish, but eventually surprised smiles were breaking out. Within moments refills were being requested and very soon the risotto was gone. Eager not to miss out, a few enthusiastic converts were now scraping the cooking pot with their plastic spoons. While the risotto had been cooking, Peter had put a chocolate cake in the oven. Not just any chocolate cake, but a mixture of white and dark cake mix, with mealworms for added texture and flavour. This now emerged

Michelin star chef Peter Gorton adds mealworms to his vegetable risotto.

from the oven resembling the famous Cornish dish Stargazy Pie in which fish heads protrude from the pastry lid, in this case however it was mealworms sticking out of the top of the cake! We cut it into portions and they were soon gone. A number of disappointed faces hovered around in case there was any more, but they were out of luck. The *Ento Café* had been a great success, and Peter and I went on to work together on many occasions.

By this time, it was early evening, the stalls had been packed away and it was time for the film for grown-ups, *Bug*. A smaller section of the audience had stayed for this but there was great curiosity. The film began and a feeling of disquiet crept across the auditorium. *Bug* depicts the decent into madness of a soldier who is convinced the military have experimented on him, infecting his body with an insect parasite. It's a graphic portrayal of self-destructive paranoia, which is both shocking and disturbing. The film drew to its tragic conclusion and the credits rolled but there was no chatter; the auditorium was silent, the audience apparently stunned, even traumatised. I was horrified! Dave looked across at me and smiled, raising his shoulders in an apologetic shrug. Well, I guess the rest of the day had gone extremely well. I made the decision to view all future films before we showed them. Dave took this on board and censored his choices of future films as all his subsequent choices were brilliant.

Dave and I organised three film festivals before he left for greener pastures, and as is the modern way, he was not replaced. It became more difficult to organise any other film events. The three that we ran had been extremely successful, attracting large audiences and offering film, food and even contemporary music at the last one. While there were no more insect film festivals, a film evening known as SciScreen did emerge. This was run by a colleague linked to the British Science Association. She thought that films based on science were often not fully appreciated as the audience sometimes lack the technical background to understand the finer points, so she would organise a scientist with expertise in the area that the film covered to talk about each film before it was shown. This meant the audience watched the film from a more informed perspective. I was asked to give talks at two of these films: David Cronenberg's film *The Fly*, a short story first published in *Playboy* magazine that deals with a scientist building a matter transporter which he tests on himself but fails to notice the fly that has joined him on the journey, and *Starship Troopers*, a novel by Robert Heinlein that deals with an intergalactic war with an insect-like adversary. These talks were great fun, and it was intriguing delving into the stories behind these films.

As you can see, the journey from talking to cubs round a campfire to discussing the science in *Starship Troopers* had been a meandering but fascinating road.

Chapter 8

Art and entomology – two cultures, both alike in dignity, in fair academia where we lay our scene

Walking down Edinburgh's Royal Mile during the Fringe Festival, I was immersed in a river of actors, performers and musicians that surged through the milling throngs of tourists who had come to enjoy the almost limitless entertainments that were on offer. Brief snatches of plays, songs and dance breakout in the crowd, while stand-up comedians and musicians vie for the attention of the public at large. Small stages were scattered along the mile and performers jostled for position. It was a colourful, chaotic staccato patchwork of performance and performers, all desperate for my attention. As I threaded my way down the Royal Mile, I accumulated a handful of flyers, thrust at me by eager hands, a brief selection of the performances available later that day.

Breaking free of the chaos, I made my way back to the comparative calm of the small theatre at the rear of the Italian delicatessen Valvona & Crolla, where entomology and opera were due to meet as *Miriam*, the opera, emerged onto the festival stage.

How did entomology and opera become bedfellows? Over the past four years I had been working with Frances M. Lynch and Karen Wimhurst, both of whom are members of the London-based music collective Electric Voice Theatre. We had been working together to produce a short opera that explored the life and works of the entomologist Miriam Rothschild.

A chaos of posters on Edinburgh's Royal Mile during the Fringe Festival.

I had met Karen at the premier of her opera that dealt with Charles Darwin's eight-year obsession with barnacles, which was first performed at Plymouth University as part of the 2009 celebrations of the bicentenary of Darwin's birth. Many of the props were specimens that had been borrowed from my laboratory museum. Karen had also asked a colleague of mine, John Spicer, and I to read through the score and check for scientific blunders. After the show I chatted with Karen over drinks in the bar and we both felt that the combination of biology and opera had more to offer.

The following year Karen made contact: she had come across a funding opportunity to produce a performance inspired by the work of female scientists. Could I suggest some candidates? I dug into the history of science and came up with Jocelyn Bell Burnell, the astrophysicist who discovered pulsars (a rotating neutron star that emits electromagnetic radiation which is observed as a pulse), but whose supervisor had won the Nobel prize for their discovery; Rosalind Franklin whose research had formed the basis for the discovery of the molecular structure of DNA by Crick and Watson, and Miriam Rothschild, a world expert on fleas and pioneer of the science of chemical ecology. After examining the list, Karen decided that 'Miriam' came top of the list.

We then researched Miriam's life. Wow, what a life she had led. She had been into so many things from setting up the UK National Parks to campaigning for gay rights. She was also a conservationist who pioneered the recovery of wild flower meadows that had been lost in the wartime expansion of farming. Born into the Rothschild banking dynasty, she was influenced by her eccentric Uncle Walter, who created a large private museum at his home in Tring in Hertfordshire. He was also famous for taming a team of zebras to pull his carriage and had used these when he visited Buckingham Palace. Both her uncle and her father Charles Nathaniel had studied butterflies as a hobby, but her father had also collected fleas, amassing the largest collection of these insects in Europe. She had also worked at Bletchley Park during the war decoding German intelligence. So she was a lady with a wide range of interests and experience.

We decided to use Miriam's own words to tell the story of her life as she had written copiously and had given a number of interviews that were now available on the internet. Karen put the words together to tell Miriam's story and wrote the music to accompany it.

The concept for the opera was to produce a performance for a single voice, Frances Lynch's, and bass clarinet, which Karen played. My role was to source material, check the science and provide suitable scientific props. I can still recall those heady days of final rehearsals in Shrewsbury where Karen and Frances crafted the delicate flow of the music and narrative that brings Miriam's life into focus. I had not heard Frances sing or a bass clarinet play before; both were a complete surprise. Frances' voice was incredibly powerful with a huge tonal range; I was transfixed. The bass clarinet was immense; it was the same height as Karen, and its deep resonant tones filled the room. Miriam Rothschild always wrote eloquently, so her words flowed seamlessly into the libretto. The sheer intensity of Frances' voice was in stark contrast to the clarinet's warm, almost sonorous sound as it danced from classical to jazz themes and back again. I came away entranced and optimistic that *Miriam* would be a success.

The first performance of the opera took place at the Royal Entomological Society's annual conference in 2013, which was hosted by the University of St Andrews in Scotland that year. At the end of the first day, delegates gathered in the long, curved bar where the Vice Chancellor of St Andrews gave a welcome speech followed by a response from the Society's President. Just as the assembled company began to dissolve into informal discussions of the day's events, a woman in a white lab coat, a purple scarf around her head and white Moon Boots walked into the crowd. She plucked something very small from the shoulder of a surprised delegate and loudly declared it a member of the Family Pulicidae (fleas).

The crowd hushed and Miriam walked around the room inspecting shoulders and sharing her delight in the biology of

Frances Lynch and Karen Whimhurst performing
the opera *Miriam* at the Edinburgh Fringe.

fleas before bursting into song accompanied by Karen's bass clarinet. The performance was a great success and Twitter went into overdrive.

Miriam has since been performed in locations across the UK and from this starting point Frances formed the company Minerva Scientific, which has gone on to develop a long series of performances that celebrate female scientists.

How does an entomologist become involved in the performing arts? It all began many years ago while I was running the first ever Build a Bug (see chapter 7) session at Plymouth City Council's annual Environmental Fair. This was a showcase for local environmental groups, and I was there on behalf of Plymouth University to inspire young people to take an interest in the natural world in general and insects in particular. Jacolly Puppet Theatre was in the room next door producing a puppet play about recycling waste with a local school while I was building insects from coffee cups and drinking straws with various groups of children. Jacolly is a two-woman team, Jack and Holly, two effervescent ladies with boundless energy who perform puppet plays with an educational message. They had been touring schools in the South West with a show, *Astra and the Waste Monster*, that dealt with recycling, and they were looking for new avenues to explore. Over lunch I talked with them about their emerging show, their hastily produced puppets and life as puppeteers. I was amazed at how quickly they had worked with the children to create a plot, characters and the stick puppets to represent them. They had just enough time for a couple of rehearsals before a performance at the end of the afternoon. I was impressed by the speed with which the show came together. and they were fascinated by the insects I was building and thought a set of insect puppets would be fantastic fun. Having had the idea placed in my head, I also began thinking this would be a wonderful idea, but such puppets would require funding and who would be crazy enough to finance such a venture? The day was a great success and we parted company with ideas floating in our imaginations.

The following week Karen Gresty, the Head of our Department's Outreach team, came into my office with a letter from the Biotechnology and Biological Sciences Research Council (BBSRC). They had set up a fund to support new innovative ways of communicating science to primary schools. The grant limit was £10,000 per project. Did we have any ideas? If ever there was a coincidence of opportunities this must be it. Nervously I suggested to Karen that perhaps a puppet play about invertebrates might fit the bill; it would certainly be innovative and definitely different.

No, wait. Surely that was a crazy idea. Why on Earth would BBSRC fund something like that? They would be looking for serious projects, surely this would be way-out of their comfort zone. On the other hand, who knows?

We hastily drew up a plan, literally on the back of an envelope. I phoned Jacolly and organised a meeting. They were as stunned as we were uncertain. But we dreamed up some costings and put together a bid. At this point we were certain that it would not be accepted, but the application would earn us a few points for effort with our Head of Department.

A month later an excited Karen walked into my office with a letter from the BBSRC. They had approved our project.

There was initial wild excitement. Wow! That is fantastic! This is going to be brilliant! I could visualise the puppets already. I phoned Jacolly again and gave them the news; they were also excited. "Brilliant, let us know when you have an outline of the play and we will start designing the puppets."

Ah, an outline, the script, the plot. Who would be doing all that? Karen and I looked at each other. Oh no… no… no, that can't be us?

Oh, my goodness. What have we done? Let's think about it over the weekend. We had been determined that this play would be based on real insects and real behaviours, not like the recent Disney animation, but that's as far as we had got.

Over the afternoon the reality of the situation sank home. Jubilation was replaced by panic. I returned home and started to think about a plot; the original idea was to create a story about garden insects. What can bring insects together in a garden? Perhaps a café? Yes, that's good. Maybe a woodlouse that runs a café? Hold on – isn't that's already way off our concept of real biology? A quick phone call to Jacolly and they confirmed that the café is a good idea; children have to relate to the overall setting.

I sketched out a loose plot and talked with Karen on Monday. We discussed ideas and threw them out. I went home and started again. In the end I spent every evening and the next weekend thinking and re-writing the text, but slowly, ever so slowly, a plot did indeed begin to take shape.

We ran it past Jacolly who were pleased with the plot but made a few helpful suggestions and, after a few more re-writes, a solid story finally emerged.

Having created the main characters, Jacolly set to work designing and building the puppets. Meanwhile, Karen and I talked to the university carpenter Brian about building the set.

Brian proved to be brilliant; he took my scribbled drawing and turned it into a real design. Over the following week the set took shape, the puppets made an initial appearance and the plot and script were polished by the team. The play would alternate between live puppets and shadow puppets on a screen centre stage. The shadow puppets allowed the plot to move quickly between different locations in the play, whereas the hand and rod puppets action always took place at the café. After many rehearsals in Yelverton village hall, we were ready for our first audience.

We invited a local school to a dress rehearsal at the University. It went reasonably well, just a few hitches to iron out, but it appeared that both the children and teachers like it. A few more tweaks and we premiered at the National Marine Aquarium in Plymouth to a couple of local primary schools.

Jacolly Puppet Theater performing Real Bugs.

Kelvin Boot, fresh from local TV and now PR director at the aquarium, introduced the play and the show burst into life. A woodlouse called Oniscus (On for short) ran a café for garden insects along with his friend the fly called Dip. An aphid dropped into the café for a quick cup of sap but left her baby behind. The baby quickly matured and produced more babies, and the café is rapidly overrun with aphids. Ants then arrived to take control and farm the aphids, kicking On and Dip out of the café. Dip then journeyed through the undergrowth on a series of adventures in a bid to get help and the show ended with the audience being showered with insect bits as the ants were expelled. The children loved it. We had made it. The craziest of ideas had become a reality and was about to hit the streets, well, schools at least.

Real Bugs became a regular attraction at any event we organised and was played at many schools, also at the Natural History Museum in London. It toured the South West for twenty years before the show ran out of steam and closed.

Tender Realm 2

The Theatre Royal in Plymouth had obtained a large grant from the Welcome Trust to commission new productions about science, and as it's the Welcome Trust, by science they mean medical science. They had begun with a call to local playwrights to produce a new work that dealt with AIDS and, as part of the research for this, the writers had visited the electron microscope at the university to see how electron micrographs were produced. The selected play had utilised images from the electron microscope as an element in the plot and, as a result, staff were invited to a reading of the selected play at the theatre's rehearsal centre. I was curious and went along; it was an interesting concept. A biologist discovers a possible cure for AIDS, but her artist boyfriend steals the electron micrograph images to make a work of art, thus making public a set of confidential data.

A clash of science and ethics, the greater good and financial gain. The reading was followed by a vigorous discussion. The play was produced and ran for a short time at the Theatre Royal. When the final production in the science series was announced, the word went round the Biology Department that the producers would like to talk to biologists. I agreed to do so and met them over a coffee on a Friday afternoon. They were Ali and Amit from the physical theatre company 'Gecko'. These guys exuded energy; they had hyperactive imaginations which jumped from one idea to another. We careered through a long, wide-ranging discussion that covered a vast swathe of biology, from evolution to death, the nature of perception and the mystery of sentience. Four cups of coffee later they asked me to be part of the science team. I went home delighted and full of enthusiasm.

The first rehearsal was arranged, and I went along to see what had been planned and how I could help. This is where perceptions are deceptive; I had assumed that I was acting in an advisory role, while Ali and Amit had assumed I was actually acting.

Rehearsals were at the theatre's rehearsal space known as TR2. It's a sleek modern building that sits on the banks of the River Plym. Rehearsals started well, with a team-building ball game that was great fun. Lots of banter and laughter; people became competitive, and we all got to know each other. Then rehearsal proper got under way. I went to take a seat and observe but was invited back onto the floor.

"NO, NO, NO!" I said. "I will just watch."

"NO, NO, NO!" Amit replied. "You are on the floor with everyone else."

I steeled myself. 'I can do this. What can go wrong?'

We walked across the room slowly changing from a happy state of mind to an unhappy one. I looked around as we traveled what now seemed a vast distance; by the time we reached the far side of the room, other members of the cast were becoming really upset and distraught. I only became acutely embarrassed; my inbuilt reserve was shouting at me that this was wrong, and I reached the other side in a state of panic rather than sadness. Over the morning we explored several other emotions and I left feeling unsure of what I was doing. Yes, what was I doing and why was I doing it? A bit of amateur dramatics in my youth is one thing but these people were real actors; I was way out of my depth.

Yet, I persevered and attended two more rehearsals in which we explored the idea of emotional research, which is the theme that the performance would focus on. TR2 would become a centre for such research, and we would each act the roles of the staff and patients. We worked out short sketches that evolved into the final performance, but each time I was told to stop acting, to be myself. I finally realised with a great sense of relief that they wanted me as a scientist, not as an actor. Nevertheless, rehearsals take time and the university gave me very little of it to play with. I told Amit and Ali that I couldn't come to all the rehearsals. "No problem," they said. "We have a small role we would like you to take on. We will give you the part at the dress rehearsal." Phew, I am off the hook but still in the production. I still had only a vague idea of what the show would look like.

I turned up at the dress rehearsal as me and awaited my part. Amit came over and I eagerly awaited my script.

"Hi Peter, this is the idea: you sit outside on those rocks and talk to the passing groups of the audience about crystals and emotions. Here is a pencil and paper, write something down. We will be back in fifteen minutes to see how you are getting on."

The world had stopped.

What!

A cold terror crept over me. NO! NO! NO! This is not supposed to happen.

"Where is the script?" I retorted.

"Use your imagination."

I was left sitting in a chair stunned and disturbed, with a pencil and a blank sheet of paper. My mind was also blank as I looked around the room ... rocks, crystals, emotions – now, really, just like that?

I could hear my heart beating.

TR2 sits by the river, and the area between the building and the river has been filled with angular grey rocks about ten centimeters across. It is a coarse version of those Zen gardens that are raked into geometric patterns, but here it's just a random grey rock field. I walked outside and sat among the rocks thinking about the emotions that were raging through me and the emotions I had been tasked to write about.

Part of the storyline is that people with emotional trauma come to TR2 to have the negative emotion dispersed. So how about if under stress people cry and in their tears we find tiny crystals. We then discover that they are the biochemical manifestations of these shed emotions. My team is researching this new phenomenon.

I ran this past Amit. "Excellent," he said. "Just trim it down a bit. You will have two minutes per group."

The production was called *Tender Realm 2*. It was a play on the letters TR2 that adorned the front of the rehearsal building. The audience turned up at the Theatre Royal in the city centre where they relaxed in the bar waiting to be called to their seats. To their surprise, they were ushered out of the theatre and onto buses that drove them out to TR2. On arrival, they were met by placard-waving protesters who shouted and heckled the audience as they entered the rehearsal space, now transformed into an emotional research centre. The protesters were objecting to the cruel and inhuman research that was being undertaken at the centre; they were an angry mob shouting and jeering as they pressed in on the disembarking audience which was quickly ushered through the chanting crowd into the reception area. Here the Centre's director greeted them and explained how the tour of his facility would work. They were then divided into three groups that were taken around the many short performances. Some were brief plays, others musicals, while some were tableaus that the audience viewed through a window.

I positioned myself on the rocks in my white lab coat and waited for the first group. I realised that I was sitting muttering to myself as I went over my script again and again, the very epitome of the mad scientist! I looked around me … the first group was approaching. Amit strode towards me in the guise of the Deputy Director. "Ah, Dr Smithers," he said. "Can you tell us a little about your research." I launched into my script and realised it was just like giving a lecture; it was going well and I glanced around the audience as I was talking and realised that my daughter's headmistress was looking back at me with an amused smile on her face. Too late now. The first group was whisked on to the next performance and I returned to gaze upon the field of rocks. The next group contained a colleague from the Marine Laboratory, but I was quite blasé by now. Calling after them as they walked away: "You can find my last paper in the recent edition of *Nature*." I was now getting cocky.

The tour ended in a small room where their guide could answer any questions that the audience may have had. Suddenly the air was filled with African drums; the walls of the rooms were slid aside to reveal the other two groups, a band and a group of dancers. There was surprise, hilarity and much dancing.

Tender Realm 2 was a surreal performance and the weirdest thing I have ever been involved in. It was a strange, testing and frequently uncomfortable experience, but I enjoyed it immensely.

At the after-show party I was talking to one of the dancers and mentioned that I had always wanted to explore the idea of insects and dance but had no idea how to go about it. To my surprise she thought that she might know someone that could be interested and would ask. A couple of days later I had a phone call: "Could I meet you at the Barbican Centre to discuss the idea further?"

We met in the theatre bar over coffee where she introduced me to Jules Laville, their in-house choreographer. We talked about my ideas, and by the time I stood up to leave, Jules said that she was keen to experiment but that we would need funding.

I agreed to explore funding from the Royal Entomological Society and BBSRC, while Jules would explore her links with the Arts Council. In the meantime, we discussed a few ideas about insect movement; Jules would talk to a group of amateur dancers that she worked with to see how they felt. This was the beginning of a great partnership. Over the following year, the dance group met one evening a week to talk and explore the choreography of entomology. I threw in lots of biology and Jules channeled this into elegant and graceful moments that the dancers then interpreted and developed. They were amazing!

The outline of a performance was beginning to take shape, but the funding was still uncertain. A letter from BBSRC arrived on my desk. I opened it tentatively, hardly daring to hope as I unfolded the single sheet of paper. YES, we had it! They had awarded us the full amount that we had asked for. The letter from the Royal Entomological Society came through a few days later with similar good news. When I rang Jules to let her know, she told me that the Arts Council had also offered us a full grant. All ten grand had come through. The show could definitely roll. Rehearsals were now more intense as it was now going to happen; the pressure had increased exponentially. We decided to broaden the show; a singer was brought in, a local composer was commissioned to write a piece for the start of show and Jules' husband Mark became involved providing a number of humorous voice-overs. The costumes were designed by the Stiltskin Theatre who also taught the dancers how to perform on stilts. The costumes were amazing, brightly coloured and padded to distort the human form into the alien shapes of the invertebrate world.

One of the opening pieces involved a group of dancers walking like a millipede. Millipedes make it look easy how they coordinate their many legs, moving them in waves of motion that pass forward along the body. A group of ten dancers has no such built-in coordination, and despite our best attempts, chaos erupted across the stage as efforts at coordinated movement generated a mass of tangled legs, tumbling bodies and lots of laughter.

While we had plenty of ideas about what to put in the show, the thing we did not have was a name. A range of titles had been thrown forward but none caught the imagination. At this time my mother was in hospital in Cumbria. After spending a few days with her, I drove to Brighton to chair the Outreach session at the Royal Entomological Society's annual conference. The session was examining ways to alter the public perception of insects. I was to give a talk about my activities, which included news about the insect dance show. While driving south, I had six hours to think about the session, my talk and the name, which I did endlessly. Somewhere on the M6 motorway the radio played some jazz and the presenter talked about the way it was syncopated. My ears pricked up. Now that's a cool word, I thought. Could I link this with insects in some way? Over the next sixty miles I ran through all the entomological terms I could remember, but no luck. The idea just did not have legs. But wait… that was it, legs were the answer. In entomology legs are termed poda. The Hexapoda are the invertebrates with six legs. So why not Syncapoda? The coordinated movement of legs or the dancing legs. We had a name at last.

Each of the rehearsals would focus on one to two aspects of the show, but how they would fit together was still unclear. After Christmas the individual pieces gelled, and we finally had an overview of the complete show. It would begin with a couple walking in a field when they noticed an insect in the grass and stopped to look at it. The couple's dance was mirrored in a film that was projected on a screen behind them. The film showed them dancing in a field. The camera zoomed down into the microcosm as we entered the insect world between the blades of grass and discovered the millipede. Caterpillars were hatching from their eggs, a praying mantis stalking the stage on stilts, huge spiders strutting before the audience and a swarm of bees waggling across stage. It

Poster for Syncapoda.

was bright, colourful, vibrant and exuded energy. We were all high as kites with optimism.

We planned to open the show over the Easter weekend, with four nights at the Barbican Theatre. There will also be entomological talks in the bar after the show, and for the gastronomically brave there were insect-based snacks available from the theatre's Thai kitchen. On the Saturday and Sunday afternoons Jacolly would perform

Syncapoda revels on the lawn at Saltram House.

Real Bugs, the puppet play for families, while on Friday evening we had an insect wrangler who would perform in the bar with a host of live invertebrates. He was a great showman as he would take an invertebrate out of a large box that sat by his chair and, while holding it in his hand, he explained its amazing biology and then put it down on the floor before taking out the next one. After fifteen minutes or so there were mantids, stick insects, scorpions, cockroaches and assorted beetles meandering around the floor of the bar. The ring of curious children was slowly contracting as they edged ever closer to get a better look, in stark contrast to the parents who were backing away towards the edge of the room; some were even beginning to climb on chairs and tables. The room was filled with an air of intense unbridled excitement and curiosity but also with a troubled anxious concern that was desperately trying to distance itself from the proceedings. The children had enjoyed the show, and despite some reservations, parents were pleased for them. All of the invertebrates were safely collected, and the bar remained invertebrate free.

Dragonfly dancers on stilts as part of a performance of
Syncapoda in Plymouth city centre.

The Guardian newspaper sent Professor Matthew Evans from Exeter University to review the performance. He wrote how much his nine-year old son had enjoyed it but then veered off on a tangent to criticise BBSRC for funding an arts project that examined insect biology, yet not funding any biological research on the topic. Still, we were reviewed in *The Guardian*.

The show went on to tour the West Country, starting at Carnglaze Caverns near Liskeard. This is a vast, man-made chamber that has excellent acoustics and has become a popular concert venue. There were also shows in Plymouth City Centre, Exeter Summer Festival, Powderham Castle near Exeter and on the lawns of the National Trust's Saltram House, which overlooks the River Pym.

Alongside the show, we ran school workshops where I would talk about insect movement and a small team of dancers would help the children move like insects.

These were brilliant, with lots of excited youngsters crawling and wriggling around the school hall while trying to be spiders,

School children learning to fly as part of the Syncapoda school workshops.

caterpillars and millipedes. Everyone enjoyed the sessions and left knowing a little more about insects and other arthropods. In June the group travelled to London where they took part in the launch of National Insect Week and grabbed a fabulous photo opportunity with celebrated TV presenter Kate Humble, a shot that made the local Plymouth newspaper, *The Herald*.

By the end of September, the costumes were in danger of becoming a biohazard as they were tricky to wash, and the dancers were dispersing to university and dance college, so the Syncapodan summer came to an end. It was a great anticlimax as it had been

The cast of Syncapoda with Kate Humble at the launch of the National Insect Week in the Natural History Museum gardens.

such an amazing project, one which I have often thought could be resurrected with the right team. I am still looking.

Syncapoda had reached a whole new audience, an audience that would never have attended a more conventional event that dealt with arthropod natural history. They came to watch the dance but left enthused about the smaller animals that we share the planet with. I also remember several of the dancers complaining to me that their partners now grumbled that Sunday afternoon walks were often interrupted by stops to look under logs and stones for invertebrates, while butterflies and dragonflies were observed at length rather than given a casual glance. It now took the dancers ages to walk anywhere in the countryside.

Ah, success. It's great when people discover something entirely new and outside of their normal experience, a new perspective on the world around them and a new appreciation of things that were previously taken for granted.

Chapter 9

Telson (the last segment of an invertebrate's body)

"If we by our art have put these wild waters in such a roar, we must allay them now"

(With apologies to Shakespeare)

The world is beautiful. It brims with life in the most extraordinary variety of forms, a beauty that I hope this book has raised your awareness of. The diversity of life on this planet is not only stunning but also awesome in its scale and complexity. Yet, in our hands the natural world has become fragile, and we have become clumsy in our care of it. Our actions have unbalanced the biosphere. We must do better.

The myriad kinds of plants, animals and simpler forms of life are almost beyond our comprehension. I could just say that there are lots of living organisms that share the world in which we live, however that would be missing the point and blurring the picture. I would like to offer readers the numbers of various species that live with us on the Earth, but sadly we don't have accurate data, just estimations. This is surprising as we know roughly the number of people living in most countries on Earth, how many stars there are in the Milky Way and the number of galaxies in the known universe, but we have only a vague idea of the number of species that live on our planet.

Surfing the Internet will find as many different estimates as you have search engine results. Scientists currently believe that there are between 1.5 and 1.9 million plant and animal species that have been discovered and documented. This means a specimen of each plant or animal had been collected, deposited in a museum somewhere in the world and someone had written a scientific paper with a detailed taxonomic description of that species. Of this huge number, 1.3 million are invertebrates (those without an internal skeleton). They are small and diverse, with about a million known species of insects alone. We may know what they look like and how to tell them apart from other similar species, where and when they were found, or perhaps a little on what the habitat is, however, we have very little information on the rest of their biology. Even though biologists have worked tirelessly to explore their lives, there are just too many of them and too few scientists working on them. There is still lots of work, especially for entomologists, to do. Vertebrates, in contrast, are generally larger animals, which are relatively easy to spot (mammals, birds, reptiles, amphibians and fish) have been studied in more depth, and their biology and distributions are known in greater detail.

Donald Rumsfeld spoke of the known knowns, known unknowns and unknown unknowns. In biology we know what has been catalogued, the known knowns, but recently there have been attempts to estimate how many species are left to discover. These are the known unknowns and you have probably already guessed that scientists around the world could not agree. However, most of them did agree that almost all of the species waiting to be discovered would be invertebrates, but estimates vary from 5 to 8 million species still to be found and recorded. The unknown unknowns in modern biology have been revealed as the bacteria and fungi. Recent research has indicated that the diversity of these groups could eclipse all the other groups combined by several orders of magnitude.

In summary, there is an awful lot of life around us, but why? What does it all do? The answer is that biodiversity is team Earth. The vast number of different species is what keeps the thin film of life, the biosphere, on the surface of the Earth stable. Each species makes use of a slightly different resource/s and processes it in a slightly different way to the other species that live in the same habitat. With many different approaches to processing that resource it is dealt with very quickly and efficiently as those species compete with each other for access. It also means that, should something change in the habitat or in the availability of that resource, some of the species will be challenged by this but there will always be some that are better able to deal with the new situation. The greater the range of expertise you have in a team, the better it can cope with new situations.

This great diversity of roles means that the plants and animals that comprise an ecosystem have evolved to utilise and recycle every possible resource within it, it is highly efficient and stable. Nothing goes to waste. However, if human activities alter the ecosystem and reduce the diversity of plants and animals, it becomes less efficient and less stable. If the loss continues and eventually too many species are lost, the system will fail. A prime example of this occurred in Maoxian County in the north-western Sichuan Province of China, where a combination of overuse of pesticides and bad orchard management led to the extinction of all local bees and other pollinators. That resulted in the fruit trees having to be pollinated by humans armed with chicken feathers! Yes, people climb into the trees and transfer pollen from one flower to another and tree to tree! A process that is both costly and inefficient, although it does produce a marketable crop in the end.

There have been many attempts to estimate the value of the pollination service that we currently have free of charge. The United Nations' Food and Agriculture Organisation produced a report that summarises many of these estimates and concluded that pollination of apples alone in the USA is worth 757 million dollars a year, while in the UK it is 81.7 million dollars. That's just the apples. The cost of pollinating all of our crops would be staggering. Is it even possible without insect pollinators to do this work?

Yet reports of declining bee populations are ringing only small alarm bells. More recent reports of a widespread decline in the abundance of all insects have generated headlines of "Insect Armageddon", but governments and funding bodies have not responded accordingly despite advice from entomologists around the world. The late E. O. Wilson (one of the greatest ecologists of our time) referred to the invertebrates as "The little engines that run the world", and indeed they are. Invertebrates are a vital cog in the maintenance of life on Earth. They are the principal recyclers of all organic materials, such as dead animal and plant remains and the faecal materials that all animals produce. They are also food for many vertebrates and could play an important role in feeding the growing human population (see chapter 6). As Joni Mitchell implies in her song 'Big Yellow Taxi', "you don't know what you've got till it's gone", we should all be mindful of this, oh so, common truth.

My old friend Ivor Kenny often complained that we live in the age of the accountant, an age where the cost of everything is known but its value is unappreciated. We need to move to a better understanding of the world where we evaluate the consequences of everything we do and consider our actions more carefully. We also need a greater appreciation of the natural world, not just for what it offers us in the form of natural services but also for its own sake. We are surrounded by incredible beauty. Let's listen to the poet W.H. Davies and take a moment to 'stand and stare'.

As a biologist, I quickly became aware of the complexity of the world in which we live, and this generated a healthy respect and awe of natural systems. Working with the arts has amplified this sense of wonder and allowed me to see the world from different perspectives, from the wide-eyed astonishment of school children to the youthful exuberance of dancers or the intensity and drama of opera. The more you look, the more you see: looking at the world through other people's eyes will reveal new wonders.

During a university field trip to Malaysia, we would take students to a small museum in Kuala Lumpur run by the University of Malaysia. Over the door was a sign that said "Everything is connected. Everything must go somewhere". If we bear this in mind as we go about our daily lives, we can move towards building a better relationship with the natural world, but this sentiment is expressed far more eloquently in the poem 'Land of our Birth' by the biologist and poet Hugh D Loxdale.

> *Confirming that this place,*
> *For which there is no price,*
> *Indeed, has worth…*
> *And that we should think well of her….*
> *And thus treat her kindly…*
> *Our unique, most lovely Earth.*

About the Author

Peter Smithers is an entomologist whose curiosity is peaked by anything with more than five legs. While he has maintained this broad interest throughout his career, it is the spiders that caught his attention and most of his research was studying the ecology of this fascinating group. He worked at the University of Plymouth as an entomologist/ecologist for forty years, finally retiring in 2013, but over that time he also worked extensively with the Royal Entomological Society where he held a number of positions, including SW regional secretary, editor of *Antenna* (the society's magazine), chair of the Insect as Food and Feed Special Interest Group and vice president. His interest in the arts has seen him collaborate with many artists and performers in order to bring the magic of the invertebrate world to a wider audience. He now lives in Bristol where he encourages his grandchildren to be excited about insects and manages a local wildflower meadow.

More books from Brambleby Books

Bee Tiger – *The Death's Head Hawkmoth through the Looking-glass*
Philip Howse
ISBN 9781908241627

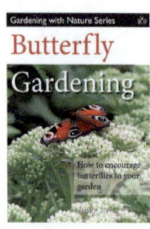

Butterfly Gardening – *How to encourage butterflies to your garden*
Jenny Steel
ISBN 9781908241436

The Butterfly Collection – *A poem for every British butterfly*
Richard Harrington
ISBN 9781908241566

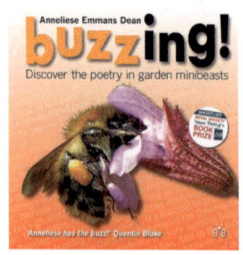

BUZZING! – *Discover the poetry in garden minibeast*
Anneliese Emmans Dean
ISBN 9781908241443

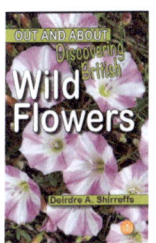

OUT AND ABOUT – *Discovering British Wild Flowers*
Deirdre A. Shirreffs
ISBN 9781908241634

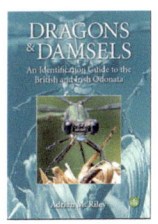

Dragons and Damsels – *An Identification Guide to the British and Irish Odonata*
Adrian M. Riley
ISBN 9781908241641

An introduction to the butterflies of Cyprus – *A photographic pocket guide to the identification of the resident species and regular immigrants*
Adrian M. Riley
ISBN 9781908241764

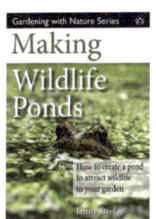

Making Wildlife Ponds – *How to create a pond to attract wildlife to your garden*
Jenny Steel
ISBN 9781908241481